Table of Contents

1. Introduction

1.1 Purpose and Scope

The purpose of this handbook is to let you become a high value engineer who masters the powerful skill stack of pump design, hydraulic calculations, financial analysis and value propositions.

The concise format aims to be a value-dense reference for both you as a student of process engineering and for you as a working professional who wishes to apply engineering knowledge, mathematics and financial analysis on real-life engineering installations.

The handbook includes step-by-step guides for your most common issues when you design and optimize pump performances. It will also guide you in hydraulic calculations and how to avoid or resolve cavitation issues.

In addition, this handbook includes a chapter on how to create impactful value propositions to any engineering investment. The idea is that you will set yourself apart from your peers in a major way, when you as an engineer are able to analyze financial impacts of technical change and know how to communicate these insights to decision makers in your business environment.

1.2 About the Author

Hey there, I'm a Chemical Process Engineer with a passion for learning, teaching and sharing my knowledge. I got my degree in Chemical Engineering from the Technical University of Denmark and the University of British Columbia, Vancouver and now live and work in Amsterdam.

During my studies I was frustrated with a lot of the heavy engineering books we had to read, because of the amount of unnecessary information packed into each chapter and the obsession with academic wording and sentence structuring. I remember thinking to myself back in the day: "There must be a simpler way of conveying these concepts and methods!"

My goal is to create valuable down-to-earth study material and handbooks for applied engineering. For this book, I have chosen the topic of hydraulic calculations and pump design based on my real-world engineering experience as it proved to be the most available skill in my toolbox coming out of university to save my clients and employers money and improve their processes.

The ability to design and optimize pumps and hydraulic systems taught in this book and the methods to present these investment opportunities to higher level management made me a very valuable asset to the organizations I worked for.

I wish for all my readers to become high-value-engineers and trusted advisors for the companies that they work for.

Thanks for purchasing this book on Pump Design & Hydraulic Calculations.
I'm excited to share my knowledge and experience with you, and I hope it will be helpful for your learning and professional development.

Humbly and gratefully,

Michael Kay Hoffmann

1.3 Overview of the Content in Main Chapters

Chapter 2 covers the fundamental concepts for hydraulic systems needed for pump design calculations, including viscosity, Newtonian and non-Newtonian fluids, laminar and turbulent pipe flow, Reynold's number, pressure drop due to various sources of friction, and the mechanical energy balance (Bernoulli's equation).

Chapter 3 will first introduce the various kinds of pumps used in the industry, along with a summarized list of benefits and limitations for each kind. It covers centrifugal pumps, positive displacement pumps, fluid properties, and pump materials. The chapter then focuses on the actual calculations that are used in the industry such as power consumption of pumps, pump and system characteristics, NPSH and operating point, and how to use MS Excel to streamline these mathematical operations.

Chapter 4 provides concise and value packed step-by-step guides for pump sizing, including choosing and designing a pump in the design phase, resolving cavitation issues in the operational phase and increasing/decreasing pump capacity in the operational phase.

Finally, Chapter 5 is a guide on how the engineer can effectively translate technical expertise into financial value. It covers the fundamentals of financial analysis needed by engineers, a guide on developing the analysis behind the value proposal, and how to effectively communicate the final project proposal.

1.4 You know why Pump-Sizing is important, right ... ?

Whether you are in the design-phase, commissioning-phase or operation-phase of a production plant, it is certain that one problem that arises (in every country, every sector, every type of plant) is pump design and performance issues.

To hammer the importance of this subject home, here is an anxiety-inducing list of risks associated with poor pump selection and sizing:

Risk	Impact
Inefficient Operation	Higher energy costs and lower overall productivity.
High Maintenance Costs	Frequent maintenance and repair needs, adding to overall operating costs.
Reduced Process Reliability	Unplanned downtime and lost productivity.
Elevated Safety Concerns	Excessive pressure and temperature, increasing the risk of equipment failure, leaks, and hazardous accidents.
Failure to Meet Operational Requirements	Not meeting required flow rates, head pressures, or other requirements leading to inoperability of the plant.
Reputational Damage	Not meeting production requirements as specified in potential contracts can damage relationships with stakeholders and clients.

Because positivity is a greater motivator for some people, let's mirror this table, and get you *pumped up* to finish and apply the knowledge of this book by listing the potential benefits of good pump selection and sizing:

Opportunity	Impact
Efficient Operation	Lower energy costs and higher overall productivity.
Lower Maintenance Costs	Reduce wear and tear damage, and reduce risk of cavitation - Ultimately providing breathing room in the maintenance budget.
Increased Process Reliability	Less unplanned downtime and more stable delivery.
Reduced Safety Concerns	More stable pressure and temperature, minimizing the risk of equipment failure, leaks, and hazardous accidents.
Meeting Operational Requirements	Ensuring smoother operability of the plant.
Good Reputation	Consistently meeting production requirements will build a good relationship with stakeholders and clients over time.

Now that you are both negatively- and positively influenced to become better at sizing pumps, let's get into the fundamental concepts that will lay the supporting substructure for the subsequent calculations and guides.

2. Fundamental Concepts for Hydraulic Systems

This chapter will provide fundamental understanding of fluid properties relevant to pump design and hydraulic calculations. After this chapter, you should have achieved the following competencies:

- ❖ Understand the concept of viscosity and differentiate between Newtonian and non-Newtonian fluids
- ❖ Distinguish between laminar and turbulent pipe flow and calculate Reynold's number
- ❖ Calculate pressure drop in piping due to friction
- ❖ Calculate the frictional coefficient (f) and units for pressure loss
- ❖ Understand pressure drop due to valves and fittings
- ❖ Understand the terms of the mechanical energy balance (Bernoulli's equation) in relation to pump calculations

2.1 Viscosity

A fluid's viscosity is its resistance to flow and is therefore a crucial factor in calculating pressure loss in the piping of the system of interest. Furthermore, very high viscosity fluids require more energy to be pumped and may even require specialized pumps to handle them.

Viscosity is typically represented in two distinct but related properties:

Dynamic Viscosity and *Kinematic* Viscosity.

Dynamic viscosity (η, eta) is a measure of the resistance of a fluid flowing under an applied force (external friction). It impacts how fast a fluid will flow through a pipe or a channel.

Dynamic viscosity is measured in force x time / area ($\frac{N*s}{m^2}$ or $\frac{kg}{m*s}$ or $Pa * s$)

The latter mentioned unit arrangement is called Pascal-second or cP (centipoise):

$$1\ cP\ =\ 1 * 10^{-3} Pa * s$$

Kinematic viscosity (ν - nu) is a measure of a fluid's resistance to flow, because of internal friction in the fluid and is independent of the fluid density.

$$\nu\ =\ \frac{\eta}{\rho} \qquad (2.1)$$

ρ (rho) is the density of the fluid. The units of the kinematic viscosity therefore are therefore in area / time and usually presented in Stoke (St) or centistokes (cSt):

$$1\ St\ =\ 100\ cSt\ =\ 1\frac{cm^2}{s} = 1 * 10^{-4}\ \frac{m^2}{s}$$

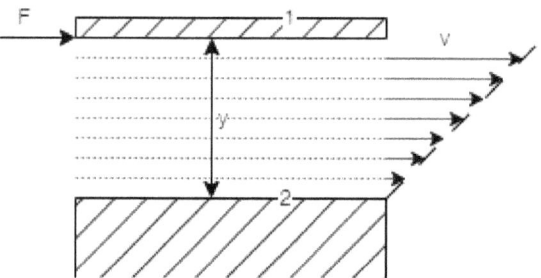

Figure 2.1.1: Shear-motion in liquids. The velocity of the layers are reduced from layer 1 to 2 due to viscosity of the liquid (energy lost due to friction).

For Newtonian Fluids such as water, figure 2.1.1 holds true as the Shear Stress ($\sigma\ =\ Force/Area$) is proportional to the speed (v) and pipe diameter (y): (Newton's Law of Viscosity)

$$\sigma = \eta * \frac{dv}{dy} = \eta * \frac{v}{y} \qquad (2.2)$$

Here the dynamic viscosity of the fluid (η) functions as a proportionality factor.

2.2 Newtonian- & Non-Newtonian Fluids

Newtonian Fluids	Non-Newtonian Fluids
Water	Ketchup
Air	Toothpaste
Most gasses	Blood
Vegetable oil	Lava
Honey	Silly Putty
Milk	Oobleck
Alcohol	Non-drip paint
Gasoline	Shaving foam
Mineral oil	Cornstarch-water mixture

Whereas Newtonian Fluids follow Newton's Law of Viscosity, where the resistance to deformation of the fluid is constant regardless of the applied force, Non-Newtonian Fluids don't adhere to this law. Their viscosity will in other words change in response to the applied stress/deformation.

The table above shows common types of fluids from each category. For this handbook we will focus on the pump design and hydraulic calculations for Newtonian Fluids.

2.3 Laminar & Turbulent Pipe Flow

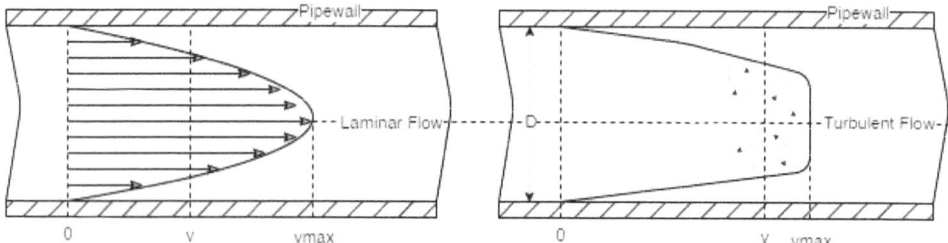

Figure 2.1.2: Flow Profiles of Laminar Flow (left) & Turbulent Flow (right).

Liquid flow in piping can occur in two different fashions.

Laminar flow means that all layers of the liquid are flowing in parallel. This leads to a velocity profile as shown in Figure 2.1.2 (left), where the max velocity is found in the center-axis of the pipe. The velocity decreases towards 0 the closer the liquid is to the inner pipe wall.

Turbulent flow means that there are whirling motions in the liquid as shown in Figure 2.1.2 (right). This leads to flattening of the velocity profile. As liquid is closer to the inner pipe wall the local flow will still behave like laminar flow. Important equations and calculation methods for pump sizing and hydraulic system design are very dependent on the type of flow happening in the pipe.

Four factors have an effect on which profile the liquid will assume: Density, flow velocity, pipe diameter and the liquids viscosity.

Before we dive into how we can quantify the flow profile of a fluid without breaking open the pipe, let's try an experiment at home:

Go to your kitchen sink and crack open the faucet a tiny bit, just enough so there is flow of water coming out. Observe the profile of the liquid stream exiting the faucet. More than likely depending on your faucet setup the flow assumes an eerily column-like shape that looks unmoving between the faucet nozzle and the bottom of your sink.

Now let's increase one parameter and keep the others constant - in other words: slowly open the faucet little by little, thereby increasing the velocity of the water.

Look closely what happens to the stream of water as you increase the velocity. The seemingly un-moving column of water will increasingly start to shake and vibrate. You are slowly moving the stream from a laminar profile to a turbulent one. At very high speeds you might even begin to see bubbles starting to form as air is getting trapped in the turbulent motions of the water.

The following subchapter dives into the math around laminar- and turbulent flow so this concept can be applied to bigger scale installations.

2.4 Reynold's Number

To predict the flow profile and thereby the relevant equations later to be used for friction calculations and eventually the pump design, engineers calculate the Reynold's Number (Re - unitless):

$$Re = \frac{\rho v D}{\eta} \quad (2.3)$$

Where ρ is the density of the fluid, v is the velocity, D is the inner pipe diameter and η is the dynamic viscosity of the fluid.

Reynold's Number and Corresponding Flow Profile		
Re < 1500	1500 < Re < 3000	Re > 3000
Laminar	Laminar, but increasingly turbulent	Turbulent

Looking at the equation above, let's reexamine the kitchen faucet experiment. We steadily increase Reynold's Number by increasing the velocity of the fluid, while keeping the other parameters constant.

The higher the Re value, the more uniform the fluid velocity through the cross-sectional area of the pipe will be because of the turbulent motions.

As mentioned previously - whether the flow profile is laminar or turbulent has a big impact in calculating the pressure drop due to friction in piping and fittings. Let's do one practice example before we move onto pressure drops due to friction:

Practice example 2.4.1: Calculating the Reynold's Number of Nitrogen

In a pipe with an inner diameter of 80 mm and a flow of nitrogen of 250 m³/h, with a viscosity of 0.016 cP and density of 1.2 kg/m³ - calculate Re and determine the flow profile of the gas.

Hints: Calculate the velocity from the volumetric flow and pipe diameter. First get the cross-sectional area $A = \pi * (D/2)^2$. Make sure the units cancel out (1 cP = 0.001 Pascal seconds = 0.001 kg/(m.s))

Answer: Re ≅ 82893 (Turbulent)

Don't worry if your result is slightly off as the exact Re is very dependent on the number of decimals used for π - as long as you are in the ballpark and can distinguish the right flow profile.

2.5 Pressure drop in piping due to friction

For laminar flow we can use Poseuille's Law to calculate pressure loss due to friction in the pipe:

$$p_f = 32 \frac{v\eta l}{D^2} \quad (2.4)$$

Where p_f is the pressure drop, v is the velocity, η is the viscosity and l is the given pipe-piece length. *Note that the pressure drop is independent of the density of the fluid, when the flow is laminar.*

For turbulent flow we can use Fanning's Equation:

$$\frac{p_f}{\rho g} = 4f \, \frac{l}{D} \cdot \frac{v^2}{2g} \quad (2.5)$$

Where g is the gravitational acceleration (approx. 9.81m/s^2) and f is the frictional coefficient, which will be detailed below.

Poiseuille's Law and Fanning's Equation will be used going forward when we are looking at Bernoulli's Equation and the Mechanical Energy Balance to define the system characteristics that lay the foundation of pump design.

2.6 The Frictional Coefficient, f

This unitless coefficient is used for pressure loss calculations in piping and fittings as a function of Reynold's Number and the relative roughness (ε/D) of the pipe material.

$$f = function(Re, \tfrac{\varepsilon}{D})$$

Roughness of different types of piping materials (ε) is the measure of the depth/peaks of the piping materials unevenness in units of length. The relative roughness is therefore a unitless number as it's divided by the diameter of the pipe in question. A table of common materials associated with piping and their roughness can be seen in appendix 6.1.

The function for the frictional coefficient f for all values of Re and ε/D is as follows (Churchill, 1977):

$$f = 2\left[(\frac{8}{Re})^{12} + \frac{1}{(A+B)^{3/2}}\right]^{1/2} \qquad (2.6)$$

The parameters A and B are defined as:

$$A = \left[2.457 \ln \frac{1}{(\frac{7}{Re})^{0.9} + 0.27\frac{\varepsilon}{D}}\right]^{16} \qquad (2.6.1)$$

$$B = (\frac{37530}{Re})^{16} \qquad (2.6.2)$$

For pipes made by materials than can be assumed to be hydraulically smooth (like glass), meaning $\frac{\varepsilon}{D} = 0$, is much simpler and is very accurate for Re > 10000:

$$f = 0.046Re^{-0.2} \qquad (2.7)$$

2.7 Units for pressure loss

One of the main concerns of doing mathematical operations on pump design and building the system characteristics is making sure the right units are used throughout the calculations, especially for the mechanical energy balances that we will explore in subchapter 2.9.

Commonly, pressures and pressure losses are described as force per unit area (Pa, Bar, mmHg i.e. Newton/Area). For pump design and system characteristics it can be preferable to measure in liquid column heights with units of length (usually meters).
This is also why the left hand side of Fanning's Equation is written as:

$$\frac{p_f}{\rho g} \text{ (units in meters instead of N/m}^2)$$

2.8 Pressure drop due to valves and fittings

The friction of fittings can be calculated using Fanning's Equation by representing different fittings as a pipe-piece by leveraging experimental values of *equivalent pipe lengths* for different types of armature.

$$\frac{p_f}{\rho g} = 4f \, \frac{l}{D} \cdot \frac{v^2}{2g} \qquad (2.8)$$

The term l/D is replaced by the unitless value of L$_e$/D, which is readily available online for different types of fittings.

$$\frac{p_f}{\rho g} = 4f \, \frac{L_e}{D} \cdot \frac{v^2}{2g} \qquad (2.9)$$

A table of common fittings and their equivalent pipe lengths (L$_e$/D) is seen in appendix 6.2. It is highly recommended to review OEM documents for specific fittings in the facility upon which you are doing your calculations to get accurate equivalent pipe lengths.

2.9 Mechanical Energy Balance (Bernoulli's Equation)

Foundational to designing hydraulic systems, designing pumps or calculating energy consumption for pumps, is the concept of an energy balance.

"Energy cannot be created or destroyed, it can only be changed from one form to another" - **Albert Einstein**

The collective expression for energy in any stream-system:

Total Energy of Stream Entering the System	+	(+) Heat Added or (-) Lost to the Environment	+	(+) Work Added By or (-) To the Enviroment	=	Total Energy of Stream Leaving the System

For our purposes and to keep this handbook relevant to pump design, let's convert this equation to a general energy balance for streams in piping.

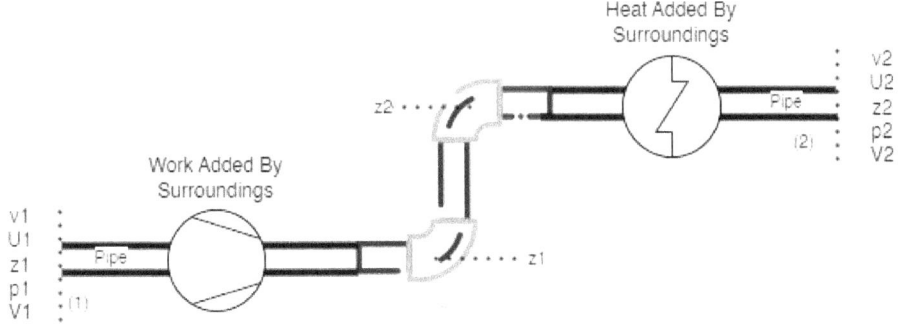

Figure 2.9.1: Process example to illustrate the concept of energy balances.

$$U_1 + \frac{1}{2}v_1^2 + z_1 g + p_1 V_1 + w + q = U_2 + \frac{1}{2}v_2^2 + z_2 g + p_2 V_2 \quad (2.10)$$

U: Internal Energy, $\frac{1}{2}v^2$: Kinetic Energy, zg: Potential Energy, pV: Newton Meter ($N/m^2 \times m^3 = Nm$), w: Work Added (+)(e.g. pumps) or Work Performed by System (-)(e.g. turbines), q: Heat Added (+) or Heat Removed (-)

Since liquids are considered incompressible as their densities change very little with temperature and pressure, we can rearrange and rewrite the energy balance for incompressible fluids: ($V_2 = V_1 = 1/\rho$)

$$(U_2 - U_1) + \frac{v_2^2 - v_1^2}{2} + (z_2 - z_1)g + \frac{p_2 - p_1}{\rho} - w - q = 0$$

If we imagine a system with no heat exchanger and negligible loss of energy to the environment (q = 0) and assume that change in internal energy is due to friction in the system, w_f, the so-called *Mechanical Energy Balance for Incompressible Fluid Flow* looks like this (with differences being denoted Δ):

$$\frac{\Delta p}{\rho} + \Delta z\, g + \frac{\Delta v^2}{2} - w + w_f = 0 \quad (2.11)$$

The terms of this form of the equation are still in J/kg.
For designing and optimizing pumps and hydraulic systems it's usually favorable to work with terms in pressure (N/m^2, Pa, Bar, etc.) or heights (m, ft, etc.)

Multiplying every term with density (ρ) to get terms in pressures:

$$\Delta p + \rho \Delta z\, g + \frac{\rho}{2}\Delta v^2 - \rho w + p_f = 0 \quad (2.12)$$

Where $p_f = \rho w_f$ (pressure loss due to friction).

By instead dividing the *Mechanical Energy Balance for Incompressible Fluid Flow* with gravitational acceleration (g), we get the equation with terms expressed in distance:

$$\frac{\Delta p}{\rho g} + \Delta z + \frac{\Delta v^2}{2g} - \frac{w}{g} + \frac{w_f}{g} = 0 \quad (2.13)$$

From these equations springs the famous *Bernoulli's Equation*:

$$\Delta p + \rho g \Delta z + \frac{\rho}{2}\Delta v^2 = 0 \quad (2.14)$$

4 assumptions drive this simple form of the energy balance:

1. Thermal Insulation - No heat is exchanged between system and environment.
   a. Technically feasible
2. No work being done by or to the system
   a. Technically feasible
3. Incompressibility of the fluid
   a. This can be assumed for liquids with negligible margins of error up almost up until the critical temperature. Gasses can only be assumed to be incompressible when the change in pressure is small compared to the absolute pressure of a given system.
4. Frictionless fluid flow
   a. Is never technically feasible, but in some cases the friction only plays a small role, giving Bernoulli's Equation multiple and impactful practical applications.

Equation 2.13 has been called the *Expanded Bernoulli's Equation* because of its usefulness as a tool in accounting for added and lost *work* in a system.

3. Pumps

This chapter focuses on the principles of pump operation and performance analysis. By the end of this chapter, you should have gained competencies in the following areas:

- ❖ Understand the operating principles of centrifugal pumps, piston pumps, and rotary positive displacement pumps
- ❖ Understand the importance of selecting the appropriate pump materials and considering fluid properties when designing and selecting a pump
- ❖ Calculate the power consumption of pumps
- ❖ Understand the concept of the operating point and how pump and system characteristics affect it
- ❖ Use MS Excel to calculate the operating point and other pump performance parameters

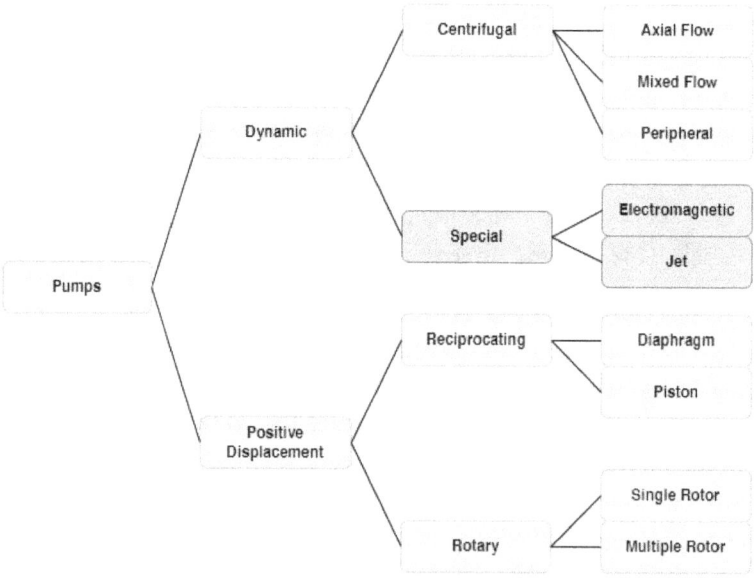

Figure 3.1: Pump types and scope of the chapter (yellow: In scope, red: Out of scope).

Liquids that need to be moved around in an industrial facility can vary a lot in physical and chemical properties. Viscosity, temperature, presence of solid particles, corrosionary properties can all put different demands on what type of pump to use.

A significant number of types of pumps exist. This handbook will cover the three most used classes of pumps: Centrifugal pumps, reciprocating- & rotary positive displacement pumps.

3.1 Centrifugal Pumps

Centrifugal pumps are most commonly occurring pumps in process engineering industries. In a pump casing, the circumference of which is shaped like a spiral, an impeller rotates at high speeds. The impeller is fitted with channels or vanes which transmit the rotary motion to the liquid. This creates a centrifugal force field. This creates a higher pressure at the periphery than at the center. This allows water to be removed from the periphery while new water is added to the center.

Here is a crude drawing to get the general understand of the flow path:

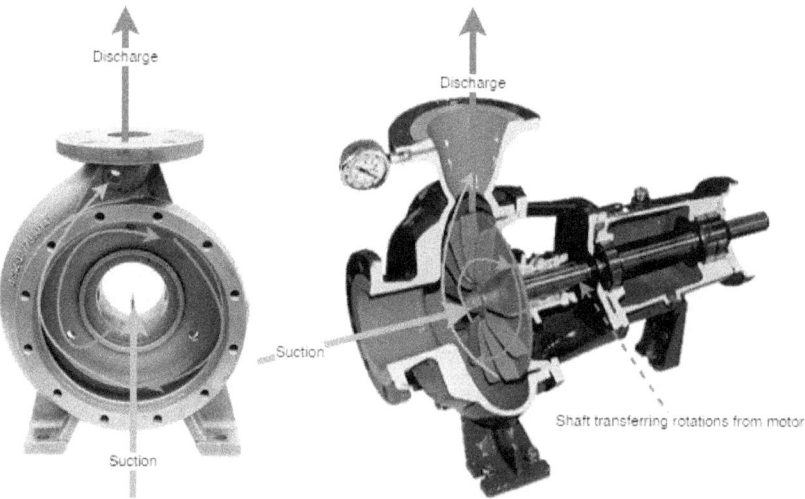

Figure 3.1.1: Flow path from the inlet-angle (left image) and from a side-view(right image).

Usually. centrifugal pumps deliver pressures up to 7 bar. Some installations will however use centrifugal pumps to reach up to 50 bar, using multiple pumps in series or through very high RPMs (revolutions per minute), usually driven through steam turbines or electric motors with gears.

In terms of volumetric flow, centrifugal pumps can have capacities in very low volumetric flow ranges to over 100 m^3 per minute. Usually the achievable pressure by a centrifugal pump will be given in head (in meters of water column), as discussed in subchapter 2.9.

Because of their simplicity, centrifugal pumps are very flexible and robust in their use compared to other types of pumps. Even though liquid is normally the medium of interest and transport by a centrifugal pumps, in sugar production facilities they are often used to transport and clean sugar beets:

By using a big centrifugal pump and a specially shaped discharge piece, sugar beets are transported (and cleaned) using centrifugal pumps. A typical size of this would be 0.5 meters diameter in the suction side pipe, 360 RPMs, power consumption around 100-200 kW and a resulting flow of 10 tonnes of beets and 60 m^3 of water per minute. Impressive application.

A centrifugal pump is constructed to transport liquids and will not work if it is filled with air. Mathematically this can be explained since the pressure increase by the pump is equal to the density of the fluid times the gravitational acceleration times the head the pump produces. At ambient temperature the density of air is around 800 times smaller than that of water, leading to a very minor pressure increase rarely able to overcome friction on the discharge side.

The upstream side (suction side) of the piping system leading to the pump therefore has to be designed to allow for liquid to flow to the pump by itself. We will dive into the concept of Net Positive Suction Head in subchapter 3.7. There exists self-priming centrifugal pumps that have complex internals to allow for the evacuation of air, but as with anything in engineering this solution comes with a price; more friction and therefore lower efficiency.

In high RPM centrifugal pumps the impeller size is made smaller compared to the suction pipe size, these types of pumps are called axial pumps is a variant of centrifugal pumps that are used to produce high volumetric flow rates against low back pressure in the discharge line.

In each subchapter describing the different types of pumps, there will be an overview of the function and deficiencies of the specific pump type to help the engineer pick the right category of pump for the specific job at hand. Here is the one for centrifugal pumps:

Important Functions and Deficiencies of Centrifugal Pumps	
Functions/Pros	**Deficiencies/Cons**
Simple construction makes it cheap and easy to manufacture in different materials	Limited pressure producing capacity
Simple maintenance a per the simple construction	Normally not self-priming.
High RPMs allows for a directly coupled motor	Usually needs check-values on the discharge side to avoid backflow once the pump is turned off.
Stable flow without pulses	Has poor efficiency with high-viscosity liquids
Piping can be blocked for short amounts of time without damaging the pump	
Smaller physical size/space requirement than other pumps with same capacity	
Can handle impure liquids	

In contrast to the centrifugal pump the following subchapter will describe piston pumps and rotary pumps, which both function with positive displacement of liquid - meaning that mechanical components of the pumps displace a well-defined volume of liquid and thereby produce an increase in pressure producing flow. This type of liquid transfer has different appliances in the process industry as seen in the following subchapter.

3.2 Positive Displacement Pumps

Piston Pumps

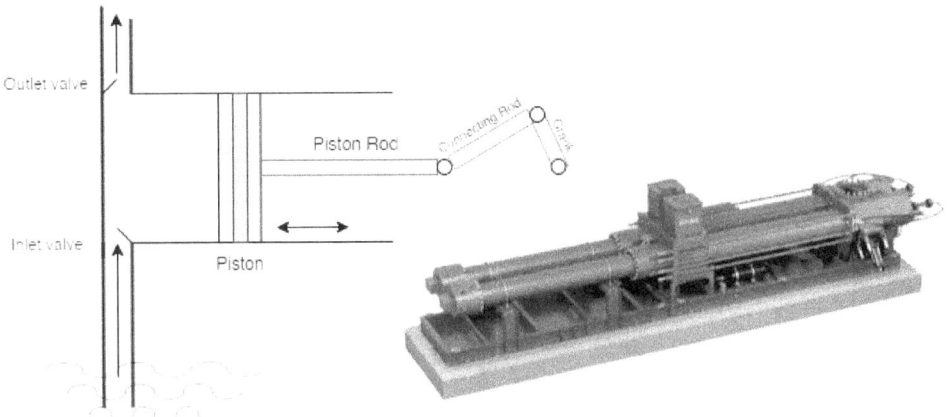

Figure 3.2.1: General function and image of piston pump.

The piston pump as seen in the above figure is usually applied for high pressures (100 bar or even more). The piston pump will move around 95 % of the exact volume of the displaced chamber per stroke. The speed is usually between 20 and 200 strokes per minute. Piston pumps have a high efficiency, usually around 85 %.

Piston pumps are usually high maintenance to keep the tightness of the piston with the cylinder, which often requires replacing of the casing/lining to keep high efficiency and minimize leakage. The pump type is therefore not suited for transporting corrosive liquids.

Membrane piston pumps exist where the only moving parts in contact with the liquid are the inlet and outlet valves. This allows this specific piston pump to handle highly corrosive liquids or sludge with abrasive solids. Membrane pistons are commonly used for chemical dosing pumps for water treatment.

Important Functions and Deficiencies of Piston Pumps	
Functions/Pros	Deficiencies/Cons
High Efficiency	Uneven, pulsing flow/pressure.
Can provide very high pressures	Takes up a lot of physical space
Good for high viscosity liquids	Purchasing price and maintenance costs are high.
Self-priming (doesn't need liquid running to the pump by itself)	
Capacity is proportional with the stroke speed and independent of pressures in the system	

The problem of pulsating flow can be vastly reduced by using duplex- or triplex piston pumps that fire with time lag.

Let's look at the other main type of positive displacement pump: **Rotary Pumps.**

Rotary Positive Displacement Pumps encompasses a large group that functions through very different principles. The 2 most common will be discussed in this subchapter.

The gear pump functions according to the heavily simplified drawing below.

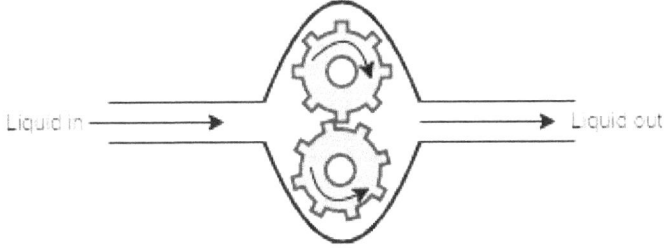

Figure 3.2.2: Simplified drawing of the flowpath in a gear pump.

One of the cogwheels is driven by a motor and moves the other cogwheel along with it. The gear pump is usually used to produce 30-40 bar, but can be constructed up to around 300 bar. The volumetric flow is almost only dependent on the speed of the gears and the displaced volume, and only very slightly dependent on the pressure of the system.

The pump works best with lubricating liquids as glycerol and oil, but can also be used to transport water. For gear pumps where both gears are driven by a motor, there is no demand on the medium's lubricating effects.

The rotan pump works with similar but more complex principles:

Figure 3.2.3: Internal view of a rotan pump.

The pump has two movable parts, a sprocket with teeth on the inner-side that has a separate driving force through a shaft through the end-cap of the pump-casing. The sprocket is in mesh with a gear that similarly to the gear pump is driven by a separate shaft. In the spaces between the teeth the liquid is moved from the suction side to the discharge side.

The rotan pump comes in many sizes with capacities between 9-3000 liters per minute. This type of pump can usually produce pressures up to around 20 bar and the efficiency can reach 80 %.

Important Functions and Deficiencies of Rotary Pumps	
Functions/Pros	Deficiencies/Cons
Stable flow with no pulses and with little to no dependency on the back pressure of the system	Prone to wear and tear from sludge-like liquids
Mid- to very high pressure production	More expensive than centrifugal pumps
Can usually pump high viscosity liquids	Demands a high amount of maintenance
Self-priming	
Physically small and doesn't take up a lot of space	

3.3 Fluid Properties & Pump Materials

When it comes to performance and longevity of pumps it is highly critical to select the correct pump material suited for operational requirements in the system of which it needs to operate. Usually pump manufacturers will provide compatibility charts that help the engineers match material-choices of the pump to the specific liquids it needs to handle. This subchapter provides a short overview of fluid properties and materials so the engineer has some background understanding when looking at these charts and in discussion with pump manufacturers.

In the grand scheme of things, pump manufacturers will know everything you need to know in this area of pump selection, but knowing general information in this field the engineer will be able to more effectively read compatibility charts and speak with the pump manufacturers.

Let's explore the 4 most common liquid properties regarding the selection of pump materials: Corrosiveness, abrasiveness, temperature and density.

Corrosiveness of the transported liquid has a big impact on pumps of the wrong material. Corrosive liquids can cause erosion, and drastically reduce the lifespan of the pump.

Corrosiveness	
Good Materials	**Bad Materials**
Stainless steel, alloy or polymer coatings	Cast iron, bronze

Abrasive liquids can cause quick wear and tear on a pump impeller and casing if these are made of improper materials.

Abrasiveness	
Good Materials	**Bad Materials**
Hardened steel, ceramic or rubber coatings	Aluminum, plastic

High temperature of liquids can cause damage and thermal expansion to pump internals if the improper material is chosen.

High Temperature	
Good Materials	**Bad Materials**
Stainless steel, alloy or ceramic	PVC, polypropylene

High density or **high specific gravity** liquids require that the pump is made of high strength materials.

High density	
Good Materials	**Bad Materials**
Cast steel or high nickel alloys	PVC, polypropylene

Below is a table for common pump materials and their properties:

Pump Material	Properties
Cast Iron	Good wear resistance, chemical resistance to mildly acidic or basic fluids, low cost
Stainless Steel	High corrosion resistance, good strength, compatible with a wide range of fluids, low contamination risk
Bronze	Good corrosion resistance, high wear resistance, compatible with seawater and brine, low friction coefficient
Plastic (e.g., PVC, PP, PE)	Good corrosion resistance, lightweight, low cost, good for handling corrosive and low viscosity fluids
Ceramic	High wear resistance, high temperature resistance, good for handling abrasive and/or highly corrosive fluids
Teflon (PTFE)	Excellent chemical resistance, low coefficient of friction, good for handling highly corrosive fluids and chemicals
Carbon Steel	High strength and durability, good for handling high pressure fluids and abrasives
Titanium	High strength, excellent corrosion resistance, good for handling seawater and highly corrosive fluids
Hastelloy	High resistance to corrosion and heat, good for handling highly acidic or alkaline fluids, pharmaceutical and chemical industries

Now, let's explore the power consumption of pumps and how to calculate it.

3.4 Power Consumption of Pumps

One of the quickest ways you can make a financial impact at the facility where you are hired as a process engineer is to calculate the power consumption of currently installed pumps and do *time to break-even* analysis on installing more efficient pumps.

Young engineers that are still early in their career (like myself) often assume that the people who designed the facility as it currently stands optimized the setup to have good power efficiency on the pumps. You will quickly learn that this is not always the case and sometimes changes have been made to the subsystem after the design phase, that changes the demand on the specific pumps. By mastering the financial aspects of engineering, you will stand-out among your peers.

This subchapter will introduce the mathematics needed to calculate power usage and pump efficiency.

The theoretical energy consumption in Joule per second (watt) can be calculated with the following formula:

$$P_t = Q\rho w \qquad (3.1)$$

Where Q is volumetric flow (m^3/s), ρ is density (kg/m^3) and w is the specific work of the pump (J/kg of transported liquid).
The specific work of the pump is obtained from the mechanical energy balance for incompressible fluids (equation 2.12):

$$w = \frac{\Delta p}{\rho} + \Delta z\, g + \frac{\Delta v^2}{2} + w_f \qquad (3.2)$$

From this it can be seen that the pump work includes the pressure increase, lift, acceleration and pressure lost to friction (not the friction loss in the pump, which will be found from calculating the pump efficiency with the actual power consumption).

As previously explored in subchapter 2.9, the relationship between head and specific pump work is as follows:

$$w = H \times g \quad (3.3)$$

The expression for the theoretical energy consumption then becomes:

$$P_t = QH\rho g \quad (3.4)$$

Now that the theoretical energy consumption can easily be calculated, let's look at the interesting part: Finding the pump efficiency. For the centrifugal pump most of the additionally consumed power comes from internal pump friction. The efficiency η_p is obtained logically from the ratio of the theoretical power consumption to the actual power consumption:

$$\eta_p = \frac{P_t}{P_e} = \frac{QH\rho g}{P_e} \quad (3.5)$$

This equation is also used to size electric motors to attach to pumps.
Typically when choosing motors, using the pump manufacturer's attached pump characteristics (pump curves) to find an estimated efficiency of the pump. This also requires building the system characteristics (system curve) to find the operating point of the pump.

Once this operating point and accompanying efficiency are known, the smallest motor that can match the actual (but calculated) power consumption is purchased.

For instance if the system characteristics leads to a theoretical power consumption of 7000 W and the pump characteristics and operating point show an efficiency around 80 %, the actual power consumption can be estimated:

$$P_e = \frac{P_t}{\eta_p} = \frac{7000W}{0.8} = 8750W$$

By searching motor catalogs to match this demand the smallest motor that can match the 8750W with a margin of error is chosen. For this example a 11kW motor would be sufficient.

These calculations become very powerful when used to design pumps, their power usage and motors or monitor the pump efficiency. The next subchapter of this handbook is key to making these calculations possible.

3.5 Pump- & System Characteristics - Operating Point

A centrifugal pump's operating characteristics are described with its pump performance curve. They typically show head, power consumption, efficiency and Net Positive Suction Head as functions of the volumetric flow rate. In the figure below, the different types of centrifugal pumps and their typical Head vs. Flow relationship is outlined.

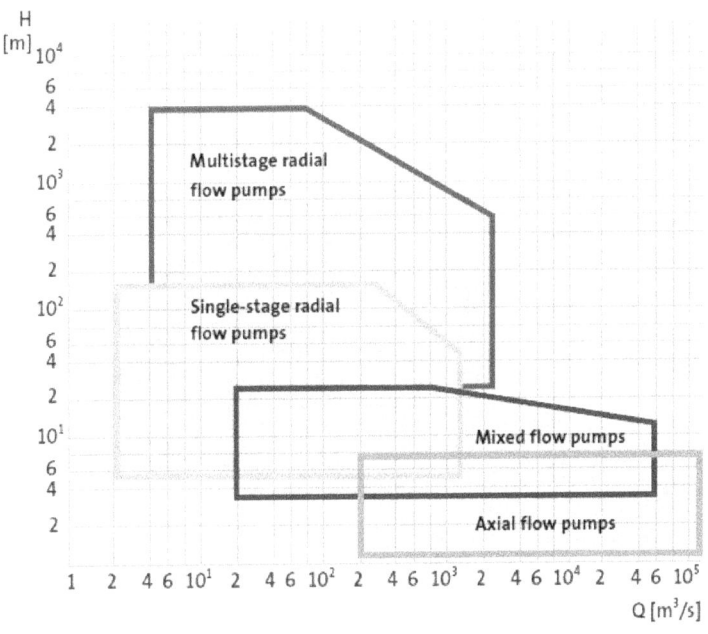

Figure 3.5.1: Areas of typical operation for pump types, Graph from Grundfos - https://www.grundfos.com/ - 28th of April 2023.

When pump curves are mentioned it is usually referring to the centrifugal pump curves, as positive displacement pumps have very simple curves, where efficiency is constant and volumetric flow is proportional to the RPMs (or strokes per minute).

Centrifugal Pump Curves Vs Positive Displacement Pump Curves		
	Centrifugal Pump Curve	Positive Displacement Pump Curve
Flow	Flow varies with pressure/Head, Higher head means lower flow.	Flow is proportional to stroke speed, and pump is known to be a volumetric pump with very predictable behavior.
RPM	One RPM unless multi speed curve (For Variable Frequency Drives).	RPM is detailed on the graph. Flow is proportional to RPM. Pressure is constant meaning pump is volumetric
Curve Shape	Sloping curve detailing drop in flow against pressure	Straight line demonstrating flow proportional to RPM with little change to flow across pressure
Efficiency	Parable-like curve with top point - Best Efficiency Point (BEP)	Efficiency is constant
Efficiency vs Viscosity	Efficiency drops off significantly with viscosity	Pump performance usually increases with viscosity
NPSH Required	NPSHr increases significantly at end of curve	NPSHr is constant

Let's look at an arbitrary pump curve from Grundfos Product catalog, and break it down:

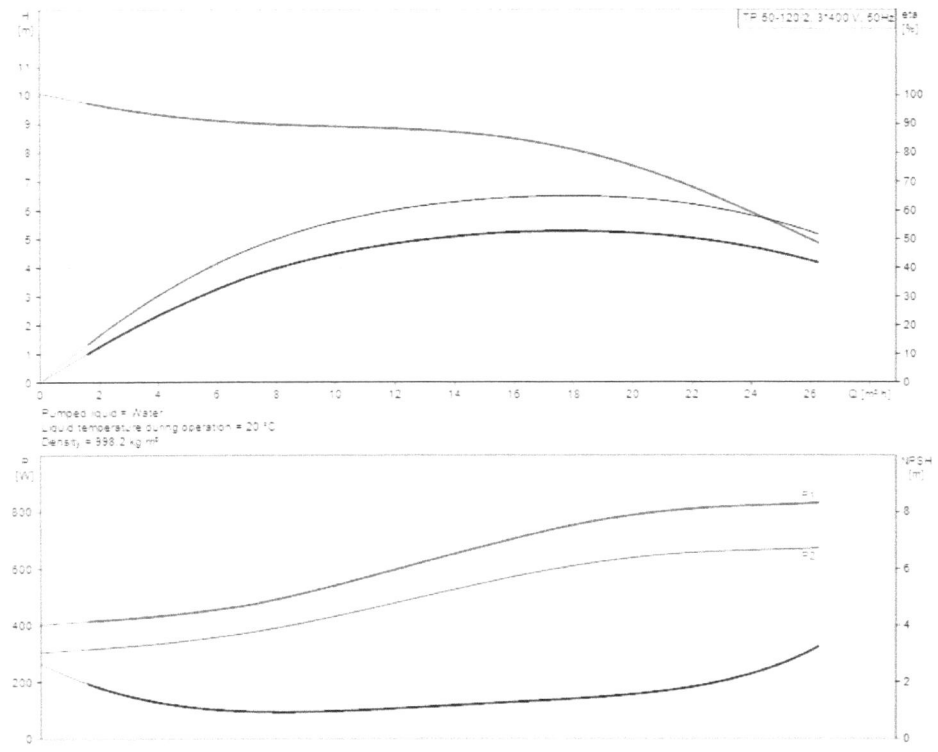

Figure 3.5.2: Performance Curve for TP50 - 188mm impeller - Grundfos Pump Catalog.

Starting from the top we have the QH-curve which describes the head (left y-axis) that the pump will produce at a given flow (x-axis).

Next line is a parable-like curve with a high vertex. This is the efficiency curve for the pump. The vertex of this curve is the Best-Efficiency-Point (BEP) and the efficiency can be read on the right y-axis. Some pump performance curves like this one will have a second efficiency curve below it, which represents the combined efficiency of the pump and the motor. This is usually included in the graph if the pump comes with an included motor-unit.

On the bottom graph (that shares the x-axis with the top graph), the first curve is P1 - the total power consumption of the pump and motor combined as a function of volumetric flow rate. If no motor comes natively with the pump, only the second QP-graph for P2 will be there. That's the power consumption of the pump itself.

Now the very bottom curve is the $NPSH_r$, but usually just denoted NPSH. This will be expanded upon in subchapter 3.6. We will be doing some example walk-throughs in this book to fully engrain the understanding of the pump curve. For now let's dive into system characteristics (System Curve) and see the important interplay between these and pump curves.

A System Curve is a QH-curve for the collective downstream and upstream factors that impact pump performance. Understanding by a real example is always good, so let's look at a case where we want to draw the system curve for a system where water is being pumped from one tank to another:

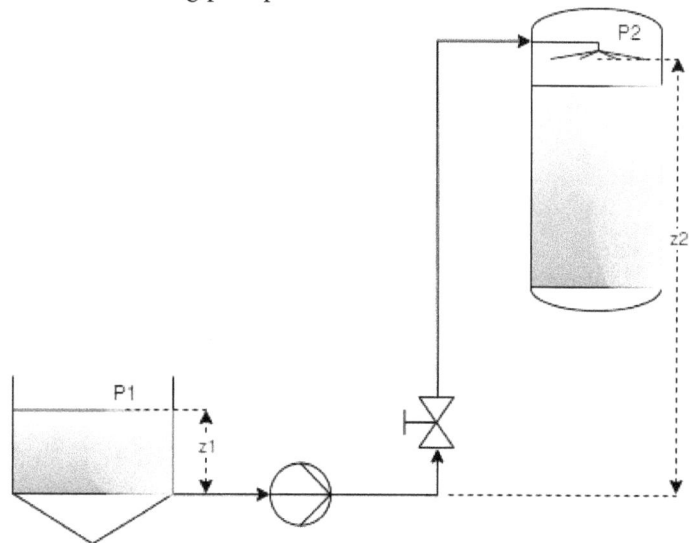

Figure 3.5.3: PFD example to illustrate system characteristics.

To create the function: H(Q), we use the mechanical energy balance derived in subchapter 3.4 (equation 3.2 & 3.3).

Using equation 3.2 where Δv is equal to v2 since v1 is insignificantly small:

$$H = \frac{\Delta p}{\rho g} + \Delta z + \frac{v^2}{2g} + 4f \frac{L_{e,total}}{D} \frac{v^2}{2g} \qquad (3.6)$$

H here represents the total pressure in meters that a pump would have to produce to lift the liquid, overcome the friction in the pipe and fittings and arrive at the second tank at pressure, P2.

The equation is missing Q, so knowing that: $Q = \frac{\pi D^2 v}{4}$ and assuming that we have strongly turbulent flow (High Re, almost constant f-factor) we can simplify the equation using the proportionality factor K for the friction in the pipe and fittings:

$$H = \frac{\Delta p}{\rho g} + \Delta z + KQ^2 \qquad (3.7)$$

From this equation the system profile will look something like this:

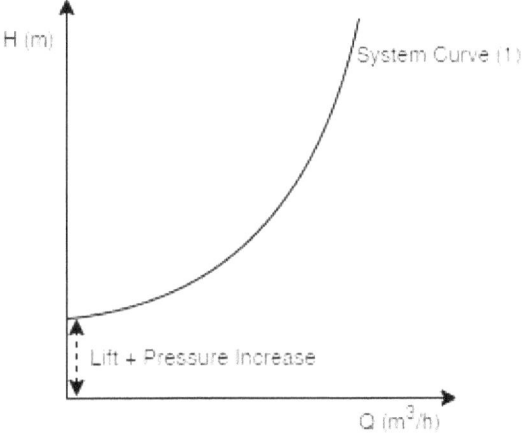

Figure 3.5.4: General system curve

Now that we have the system curve, here is where the magic happens. We overlay the graph with the pump curve for the pump that we are considering (if designing a new system) or for the current pump if we are studying the current setup.

Overlaying a pump curve will then look something like this:

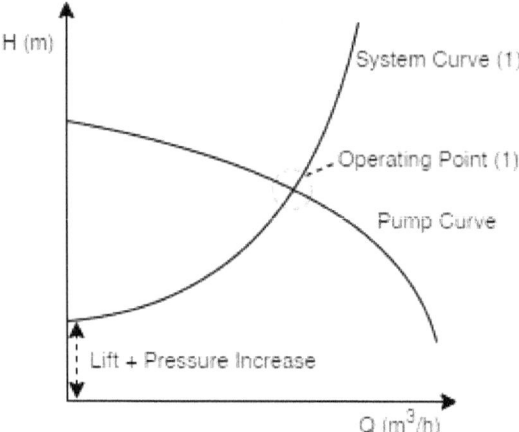

Figure 3.5.5: General system curve and pump curve showing intersecting operation point.

Now we have the operating point which is the actual head the pump will produce and the resulting volumetric flow we will have at the spray nozzle of the discharge tank.

What might happen to the flowrate and head produced by the pump if we close the gate valve on the discharge side of the pump a little bit? The pump curve will remain the same, but we will shift the system curve. Closing the valve a little increase the collective friction factor K and thereby shifts the operating point like this (2):

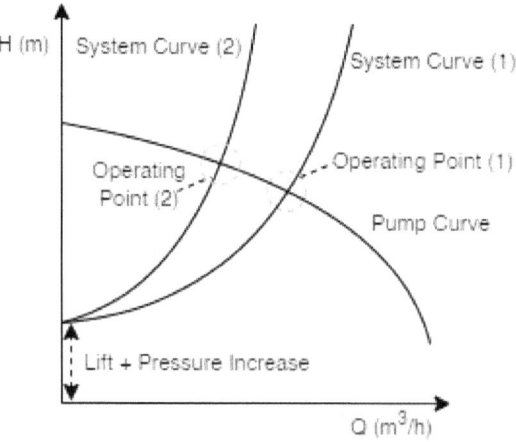

Figure 3.5.5: Showing shift in system curve due to an increase in discharge side friction.

The resulting flow rate can thereby be controlled by shifting the system curve. As a result of this mechanism orifice plates are widely introduced in the industry to add friction downstream and thereby decrease the volumetric flow rate or to operate closer to the BEP of the pump curve.

In the following chapter we will explore a real-life example of finding operational points and solving common process engineering issues like calculating power consumption or restricting flow rate and how these can be calculated using software like MS Excel.

3.6 Using MS Excel to Find the Operating Point

One of your colleagues bought a new pump for your production line and it's your job to check the maximum flow and the power consumption of the pump once implemented.

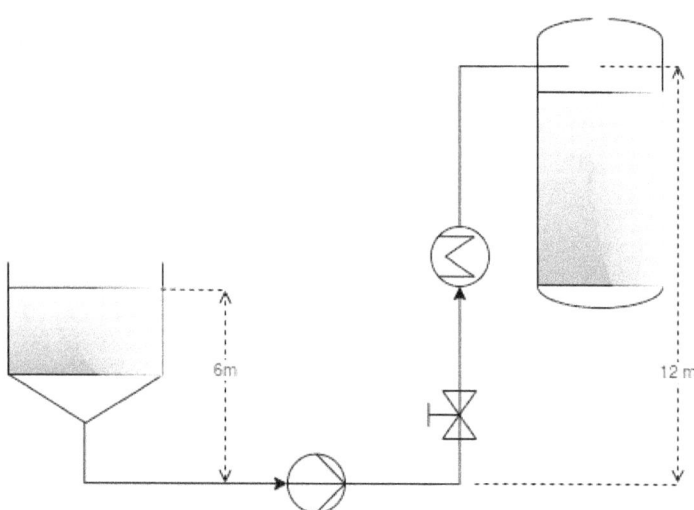

Figure 3.6.1: PFD of the given system.

The total pipeline from the first tank to the second is 25 meters, 50 mm diameter, stainless steel and with a roughness of 0.05 mm. The combined pressure loss of the fittings in the line is 30 velocity heights $(= 30\frac{v^2}{2g})$.

The density of the transported product is 1000 kg/m³ with a viscosity of 1 cP = 10^{-3} Pascal seconds.

Setting up the equation for the total head needed to move the fluid at different flow rates (using equation 3.6):

$$H = \frac{\Delta p}{\rho g} + \Delta z + \frac{v^2}{2g} + \frac{p_{f,piping}}{\rho g} + \frac{p_{f,fittings}}{\rho g}$$

Inserting the following known values and relationships:

$$\Delta p = 0, \ \Delta z = 6, \ \frac{p_{f,piping}}{\rho g} = 4f\frac{L}{D}\frac{v^2}{2g}, \ \frac{p_{f,armature}}{\rho g} = 30\frac{v^2}{2g}$$

$$v = \frac{Q/3600}{D^2\pi/4}, \quad Re = \rho v D/\eta$$

For f (friction factor) use equation 2.6 or the Moody Diagram[1]

Leveraging the computational power of MS Excel to find a reproducible method to solve these tasks is highly beneficial for reproducibility and iterations. As an Engineer you probably like to automate things.

Fixed Variables	Pipe Roughness	D	ρ	η	Length	z2-z1	p2-p1	Fittings friction (v heights)
Value	0,0005	0,05	1000	0,001	25	6	0	30
Units	m	m	kg/m3	Pa*S	m	m	Pa	-

Q	v = (Q/3600)/(D^2*PI/4)	Re = ρvD/η	f	v^2/2g	H
m3/h	m/s	-	-	m	m
0	0,000	0			6,00
12	1,698	84883	0,0057	0,147	12,22
14	1,981	99030	0,0056	0,200	14,43
16	2,264	113177	0,0055	0,261	16,96
18	2,546	127324	0,0055	0,330	19,87
20	2,829	141471	0,0054	0,408	23,04
22	3,112	155618	0,0054	0,493	26,62
24	3,395	169765	0,0054	0,587	30,54
26	3,678	183912	0,0053	0,689	34,66
28	3,961	198059	0,0053	0,799	39,24

Figure 3.6.2: Example of how the known data and calculations can be entered in MS Excel.

Select the column containing the Q-values, hold CTRL and select the H-values and then insert an XY-scatter plot:

---

[1] https://www.engineeringtoolbox.com/moody-diagram-d_618.html

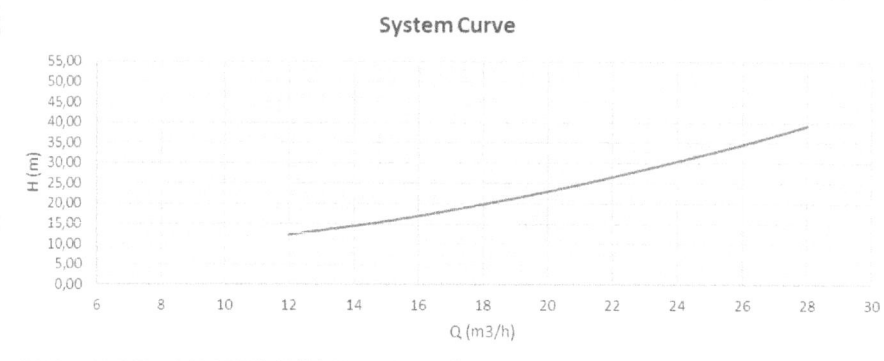

Figure x: System Curve Scatter Plot

Add appropriate labels and gridlines for readability.

Now that the system curve has been made let's look at the pump that your colleague has purchased. Luckily it was a pump made by Grundfos who has excellent online pump documentation (even interactive pump curves).

Typically, you can find pump curves in the pump documentation included when receiving the pump. Otherwise most manufacturers have them available online. Worst case, you call or email the manufacturer and request the documentation

Google the pump name: NK 32-160.1/177 and arrive at the following product page and associated pump curves:

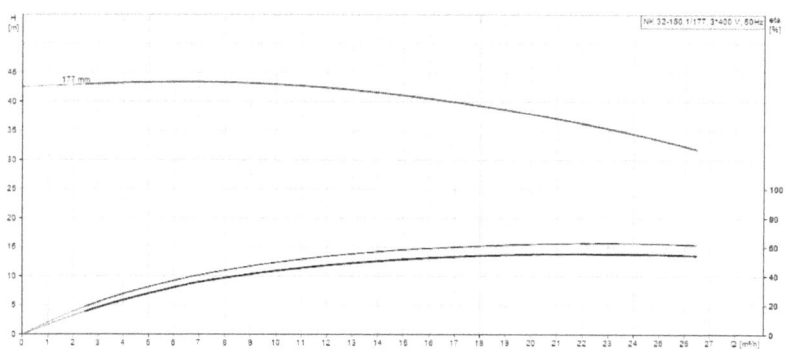

Figure 3.6.3: Pump Curve from Grundfos: (https://product-selection.grundfos.com/).

Using the interactivity of the online pump curve find Q-H pairs to add to your Excel sheet using the mouse:

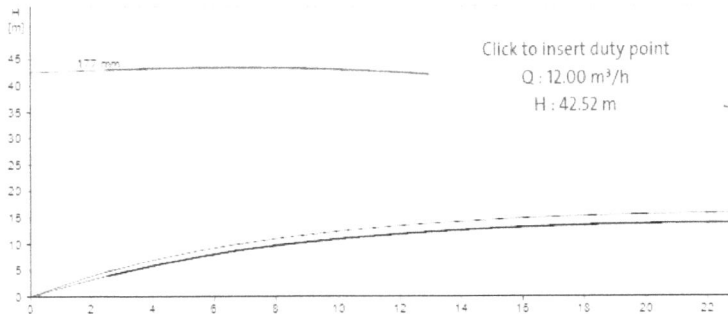

Figure 3.6.4: Using the pointer to find points along the QH-curve for NK32-160.1/177.

If you are looking at pump curves for non-Grundfos pumps, it is totally fine just using grid lines on the graphs to collect QH-datapoint pairs and plot them into Excel. Plotting the QH-data points for the pump curve into Excel:

Q	H (System)	H (Pump)
m3/h	m	m
0	6,00	
12	12,22	42,5
14	14,43	42
16	16,96	41
18	19,87	38
20	23,04	37,8
22	26,62	36
24	30,54	34,5
26	34,66	32,5
28	39,24	

Figure 3.6.5: Screenshot of MS Excel table including flow, H(system) and H(pump).

Add the Pump Curve to your System Curve scatter plot:

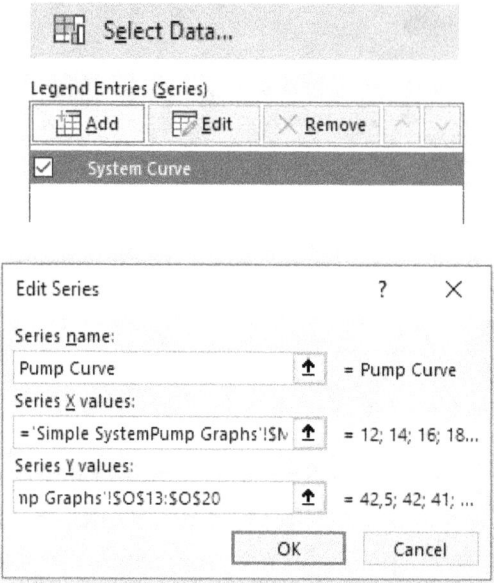

Figure 3.6.6: Adding Pump Curve to already existing System Curve Scatter Plot

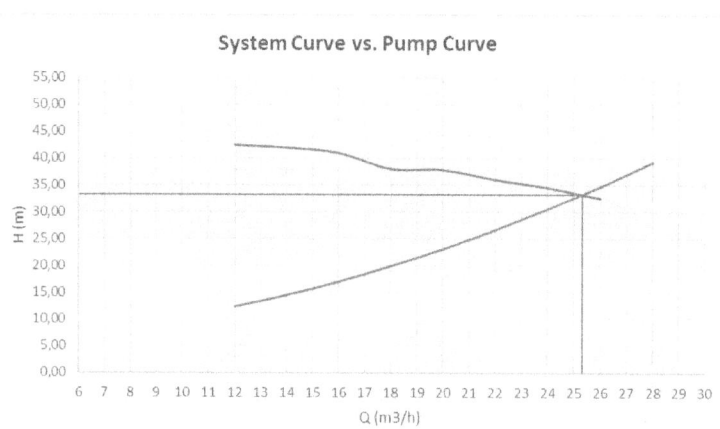

Figure 3.6.7: Combined System Curve and Pump Curve Scatter plot revealing the operating point.

By adding gridlines, you find the operating point to be at 25.2 m³/h and 33 meters of head.

You can use Excel to get a very exact intersection (operating point) between the two curves using several methods including trendlines and goal seek. For using trendlines the approach is as follows:

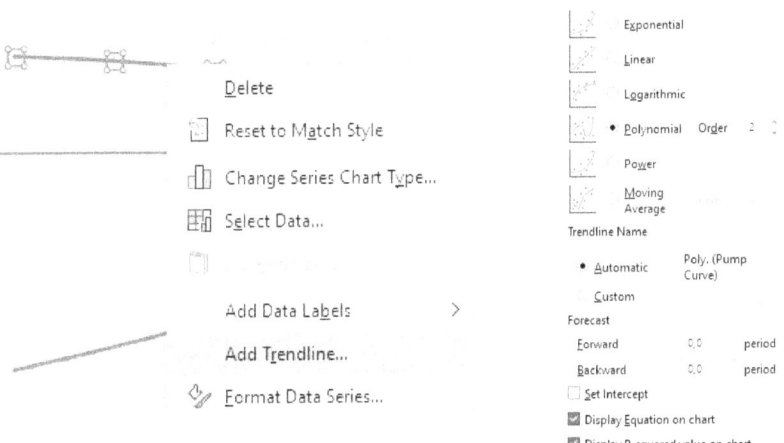

Figure 3.6.8: Adding trendlines to curves with equations

Add trendlines for both curves and use a fit that has an R^2-value close to 1. For this example both fits are second order polynomials:

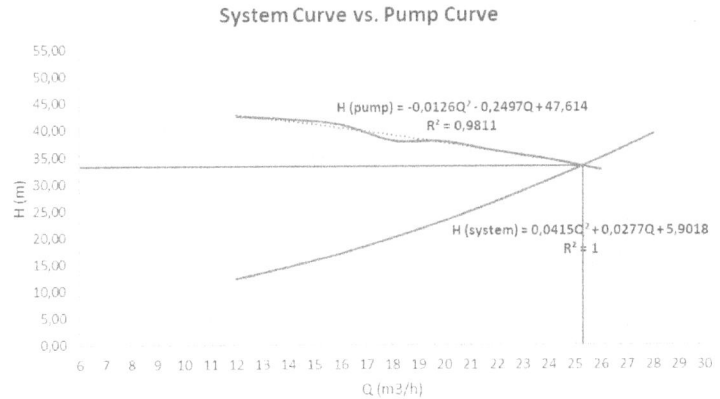

System Curve vs. Pump Curve

$H\,(pump) = -0,0126Q^2 - 0,2497Q + 47,614$
$R^2 = 0,9811$

$H\,(system) = 0,0415Q^2 + 0,0277Q + 5,9018$
$R^2 = 1$

Figure 3.6.9: System Curve vs. Pump Curve with trendlines

From this the approximate equations for both curves are obtained:

$$H\,(system)\ =\ 0.0415Q^2\ +\ 0.0277Q\ +\ 5.9018$$
$$H\,(pump)\ =\ -\,0.0126Q^2\ -\ 0.2497Q\ +\ 47.614$$

We find the intersection by setting the equations equal to each other and solving for Q:

$$H(system)\ =\ H(pump),\ solve\ for\ Q:$$
$$Q\ =\ 25.32$$

Inserting Q = 25.32 in one of the formulas to get H:

$$H\ =\ 0.0415\times 25.32^2\ +\ 0.0277\times 25.32\ +\ 5.9018\ =\ 33.2$$

Now that you have the actual flow and head that the pump will produce in the system you calculate the theoretical power consumption of the pump:

$$P_t\ =\ QH\rho g\ =\ \frac{25.32}{3600}\times 32.3\times 1000\times 9.81\ =\ 2290\ W$$

From the pump curve you find the pump efficiency at the given operating point:

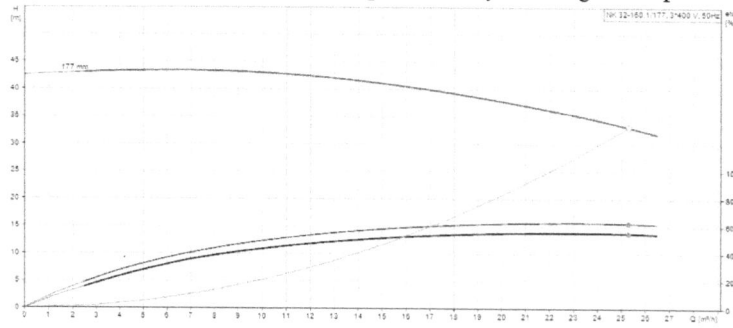

Figure 3.6.10: Pump efficiency found on Grundfos Pump Curve by inserting the operation point.

The pump efficiency at the operating point is around 62.5 %

The effective power consumption of the pump is then calculated from eq. 3.5:

$$P_e = \frac{P_t}{\eta_p} = \frac{2290\ W}{0.625} = 3664\ W$$

Some pump curves will include the effective power consumption point so this last calculation step can sometimes be skipped and just read from the graph.

In this example the maximum flow rate was found to be 25.32 m³/h given the system conditions. Let's imagine that the given process requires that only 20 m³ of liquid flow from one tank to the other per hour.

We will explore three ways of reducing the flow (aside from finding a new pump or smaller impeller since our colleague just bought this pump) and calculate the resulting power consumption of the pump:

1. Slightly closing the valve on the discharge side of the pump
2. Apply a VFD (variable frequency drive) to change the RPM and thereby the capacity of the pump

1) Closing the valve until a flow rate of 20 m³/h is reached (shifting the system curve as discussed in subchapter 3.5, will result in 37.9 meters of head being produced by the pump with an efficiency of 62.3 % as seen per the pump curve below:

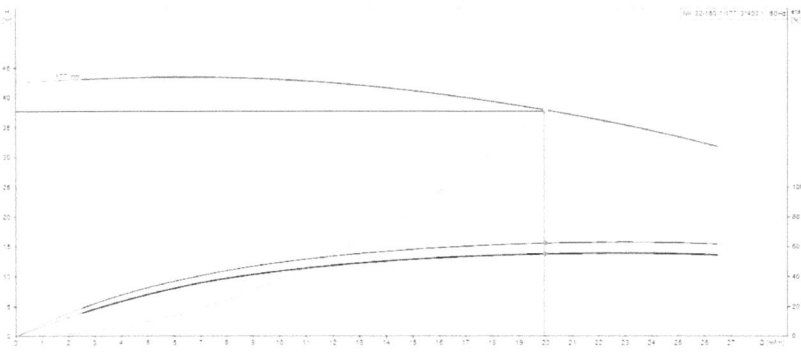

Figure 3.6.11: Pump Curve with a new Operating point at Q = 20 m³/h.

$$P_t = QH\rho g$$

$$P_e = \frac{QH\rho g}{\eta_p} = \frac{(20/3600)\times37.9\times1000\times9.81}{0.623} = 3315\ W$$

2) By changing the RPM of the pump with a VFD, the pump curve is shifted but the system curve remains the same. To calculate the reduction in frequency delivered to the pump in % needed to reduce the flow rate from 25.32 m³/h to 20 m³/h, the affinity laws for centrifugal pumps are used.

These laws state the following relationship:

$$\frac{Q_1}{Q_2} = \frac{n_1}{n_2}\frac{D_1}{D_2} \qquad (3.8)$$

$$\frac{H_1}{H_2} = \left(\frac{n_1}{n_2}\right)^2\left(\frac{D_1}{D_2}\right)^2 \qquad (3.9)$$

$$\frac{P_1}{P_2} = \left(\frac{n_1}{n_2}\right)^3\left(\frac{D_1}{D_2}\right)^3 \qquad (3.10)$$

Where n are RPMs and D are impeller diameters.

Since the system curve remains unchanged and the goal with the VFD is to be at a flow rate of 20m³/h, the resulting head the pump has to produce is found from the graph:

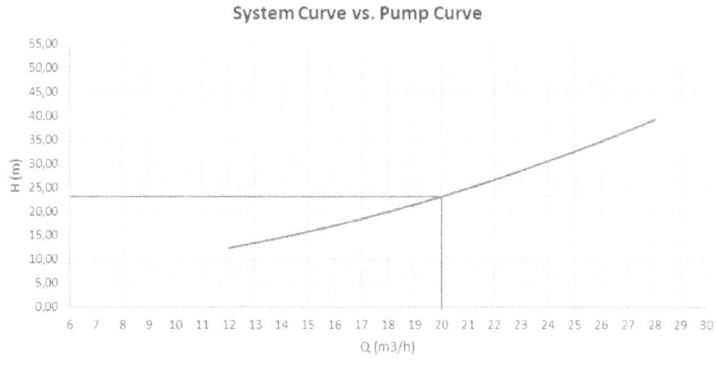

Figure 3.6.12: System Curve unchanged and $Q_2 = 20$ m³/h, $H_2 = 23$ m.

For this example the impeller diameter is unchanged ($D_1 = D_2$):

$$\frac{Q_1}{Q_2} = \frac{n_1}{n_2}$$

$$\frac{H_1}{H_2} = \left(\frac{n_1}{n_2}\right)^2$$

Therefore the following must be true ($n_2 = 100\%$ i.e. no reduction in RPM):

$$\frac{H_1}{H_2} = \left(\frac{Q_1}{Q_2}\right)^2 = \frac{n_1}{n_2}$$

$$\frac{H_1}{23} = \left(\frac{Q_1}{20}\right)^2 = \frac{n_1}{100\%}$$

Reducing and rearranging to find an expression for H_1 as a function of Q_1:

$$H_1 = \frac{H_2}{Q_2^{\,2}} Q_1^{\,2} = \frac{23}{20^2} Q_1^{\,2} = 0.0575 Q_1^{\,2}$$

Adding the function to the Excel calculation and plotting (the affinity curve), to find the intersection with the original pump curve:

Q	H (System)	H (Pump)	H (Affinty)
m3/h	m	m	m
0	6,00		
12	12,22	42,5	11,27
14	14,43	42	14,72
16	16,96	41	18,63
18	19,87	38	23
20	23,04	37,8	27,83
22	26,62	36	33,12
24	30,54	34,5	38,87
26	34,66	32,5	45,08
28	39,24		

Figure 3.6.13: Screenshot of MS Excel table including flow, H(system), H(pump) and H(affinity).

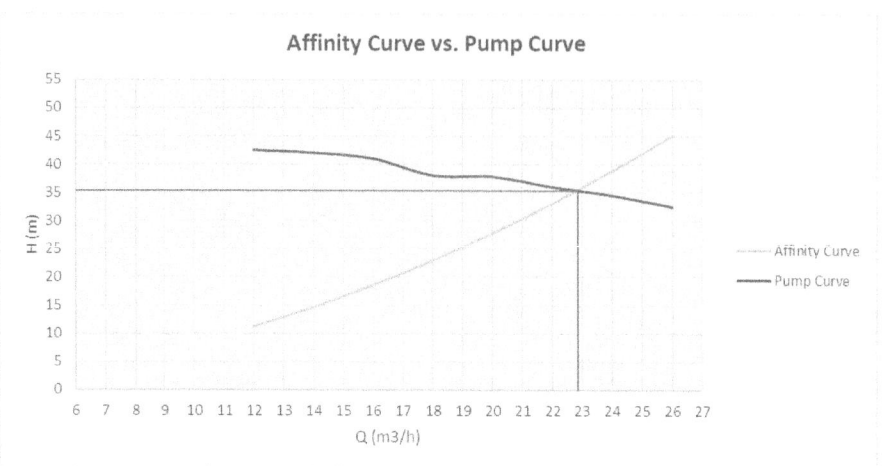

Figure 3.6.14: Intersection between affinity curve and original pump curve.

The intersection Q_1, H_1 is read to be: (22.9, 35.05). These values are inserted in the previous formula:

$$\frac{Q_1}{Q_2} = \frac{n_1}{n_2}$$

$$\frac{20}{22.9} = \frac{n_1}{100\%}, \; n_1 = \frac{20}{22.9} * 100 = 87.3\%$$

In order to achieve the flow of 20 m³/h, the VFD has to reduce the RPMs to 87.3 % of the max RPM.

Assuming an insignificant change in pump efficiency the resulting power consumption of the RPM reduced pump is calculated:

$$P_e = \frac{QH\rho g}{\eta_P} = \frac{(20/3600)\times 23 \times 1000 \times 9.81}{0.623} = 2012 \; W$$

$$3315 \; W - 2012 \; W = 1303 \; W$$

That's a total reduction in power consumption of 1.3 kW compared to the previous solution closing the valve.

To calculate the complete reduction in power consumption, the loss of energy in the VFD has to be considered. Usually around 5 % of the VFDs effect is a good estimation for the loss. For instance if a 4kW VFD was used for this pump, the loss would amount to around 0.2 kW, making the total saved amount of power to be 1.1 kW.

The engineer would then do financial analysis on the purchase price of the VFD and the man-hours to install it against the energy prices and calculate the payback period of the investment. Different companies have different time-to-breakeven standards for investments, so this would have to be compared with the "free" and instant solution of just closing the valve slightly. More on this in chapter 5.

3.7 Net Positive Suction Head - Avoiding Cavitation

To achieve smooth operation one of the key things that the process engineer is responsible for is to avoid pressures dropping so low anywhere in the pumps and in the suction lines that gas-bubbles are formed.

When liquids in pipelines evaporate due to the pressure dropping below the vapor pressure this phenomenon is called **cavitation**. Air bubbles created under these circumstances can violently implode once the pressure again rises above the vapor pressure.

When cavitation happens to a lesser degree this causes vibration that makes a crackling sound. Minor cavitation usually reduces the lifespan of the pump internals and reduces the efficiency of the pump. If the cavitation causes enough evaporation it can even shut-off the pump due to lack of liquid. For certain processes, a pump-shutoff can cause severe issues without proper safeguards.

For almost all processes the highest risk of cavitation is in the suction side, particularly pump inlet where the pressure is at its lowest point.

Available Net Positive Suction Head ($NPSH_a$) is a measure of how much head (pressure) is available on the suction side of a pump. It's a function of the absolute pressure on the liquid surface on the tank from where liquid is pumped (H_a), the height difference between the liquid surface and the pump inlet(H_h), pressure loss due to friction in the suction pipe (H_f) and the vapor pressure of the flowing liquid (H_v).

$$NPSH_a = H_a - H_h - H_f - H_v \qquad (3.11)$$

H_a, H_f and H_f are all positive values, while H_h is negative if the pump is positioned lower than the liquid surface of the tank being transported from. All terms are usually in units of meters liquid column.

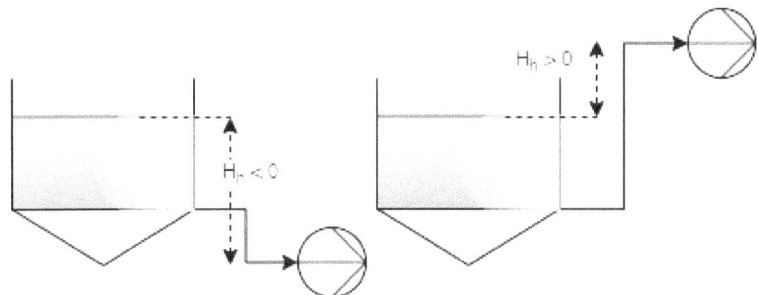

Figure 3.7.1: Visual representation of the H_h part of $NPSH_a$

The $NPSH_a$ is to be compared to the $NPSH_r$ (Required NPSH).

The required NPSH is an experimental value and is different for each pump due to the pump's specific shape and internals. $NPSH_r$ is a function of the flow rate, and this relationship is provided by the pump manufacturer and is usually also drawn on the pump curve as seen in figure 3.5.2 (subchapter 3.5).

The condition for cavitation-free operation of the pump is that $NPSH_a$ is more than the $NPSH_r$ under all operating conditions.

Let's look at an example from the design phase of a project, where it has to be calculated how low the pump inlet pipe has to be placed in order to operate without the risk of cavitation:

Condensate (boiling water at 1 bar) needs to be transported from a collection tank at 20 m³/h. The suction side of the piping is very short meaning that the pipe friction is negligible. The only friction on the suction side comes from a fully opened globe valve. The OEM of this valve has informed that its friction is equivalent to 6 velocity heights. The suction pipe from the tank nozzle to the pump has an inner diameter of 50mm (0.05m). From the Pump Curve supplied by the OEM, the NPSH$_r$ is 2 meters at 20 m³/h.

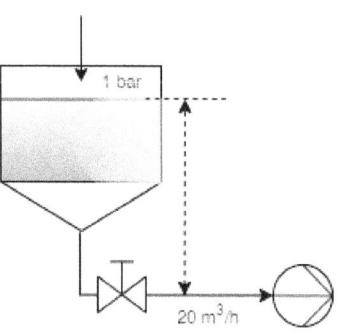

It is now the process engineer's job to calculate how low to position the pump to ensure cavitation-free operation.

The goal is to ensure that NPSH$_a$ > NPSH$_r$. Since the liquid is at its boiling point/vapor pressure (H$_a$=H$_v$) in the condensate collection tank equation 3.6.1 becomes:

$$NPSH_a = -H_h - H_f > NPSH_r$$

H$_f$ is calculated from Fanning's equation (equation 2.8):

$$H_f = \frac{p_f}{\rho g} = 6\frac{v^2}{2g}$$

$$v = \frac{Q}{A} = \frac{20/3600}{\pi 0.05^2/4} = 2.83 \; m/s$$

$$H_f = 6\frac{2.83^2}{2\times9.81} = 2.5 \; m$$

Knowing H_f:

$$- H_h > NPSH_r + H_f$$
$$- H_h > 2\,m + 2.5\,m$$
$$- H_h > 4.5\,m$$

The process engineer concludes that the pump needs to be positioned at least 4.5 meters below the surface level of the condensate tank.

It is of course best practice to think about cavitation-free operation in the design-phase rather than having to do it once the problems occur, but the problems and impacts of cavitation are seen happening on many plants many years after the design phase.

A classic example is the operation of a plant that has existed for a long time. Suddenly a centrifugal pump starts showing early signs of wear and tear even though the next scheduled maintenance is far away. As an engineer looking for the lowest effort solution, you should first check the terms from the NPSH equation that are most likely to have shifted since the design phase. For the example above the surface level of the liquid in the collection tank is critical. It might just be that the tank is now being operated in a way where the liquid surface level is 0.5 meters lower than intended. The fix here is easy.

It could also be that the friction in the piping or valve has increased due to fouling and it just needs a good cleaning, thereby lowering H_f.

Utilizing the methods of the book thus far, the following chapter will have step-by-step guides for different pump design and optimization scenarios.

4. Pump Sizing Walk-Throughs

This chapter consists of four Step-by-Step guides for process engineers to follow for different scenarios that involve the understanding of Pump Design and Hydraulic Calculations. With this guide you will be able to:

- ❖ Choose and design a pump in the design phase for a new system
- ❖ Resolve cavitation issues on an already operational system
- ❖ Decrease the pump capacity of an already operational system to meet new operational requirements
- ❖ Increase the pump capacity of an already operational system to meet new operational requirements

The following step-by-step guides will refer back to subchapters in this book that are relevant for the specific steps.

4.1 Step-by-Step - Choosing and Designing a Pump in the Design Phase

By Design Phase it is meant that the System is not yet built and iterations can be made to things like instruments/fittings/piping can be made easily.

Step 1: Determine operational & system requirements from the initial PFDs/P&IDs and build your System Characteristics Curve (subchapter 3.5).

Step 2: Determine the Pump Type to match the operational requirements and fluid properties of the system (subchapters 3.1-3.3).

Step 3: Check the Upstream Requirements and the available NPSH (subchapter 3.7).

Step 4: Choose size of pump based on Step 1-3, and choose materials of the pump (impeller, casing, seals, bearings) that are appropriate to the fluid properties. (subchapter 3.3)

Step 5: Verify the chosen pump by finding the operational point, comparing $NPSH_r$ at operational extremes (lowest- to highest flow rates) with $NPSH_a$, looking at power consumption versus alternative solutions (subchapters 3.4-3.6).

Step 6: Verify design and pump type with simulation tools such as AFT Fathom or AspenPlus to determine if the manual verification calculations were done correctly.

Step 7: Take precautionary measures to avoid cavitation. For instance installing level sensors on suction side vessels, to ensure that $NPSH_a$ does not go below $NPSH_r$ due to the liquid surface level dropping too low (subchapter 3.7).

4.2 Step-by-Step - Resolve Cavitation Issues in the Operational Phase

By Operational Phase it is meant that the pump system is already built and is in operation. Several factors can lead to cavitation starting to occur even if the system was previously designed to handle it.

Step 1: Determine if cavitation is the issue. Perhaps noise and vibration has been observed or the capacity of the pump has been observed to deteriorate faster than expected or higher power consumption by the pump is noticed.

Step 2: Determine the possible reasons (and solutions) for the pump system changing from cavitation-free operation to showing signs of cavitation. The most common reasons for cavitation developing in a normally functioning system are:

❖ **Changes in operating conditions**

➢ Example 1: Downstream valve has been opened slightly more and the resulting higher flow rate is accompanied by a much higher $NPSH_r$ as per the pump curve.
➢ Solution: Return valve to original position.
➢ Example 2: Upstream valve has been closed slightly reducing $NPSH_a$.
➢ Solution: Open valve to original position
➢ Example 3: Operation with a lower liquid level of the upstream vessel reducing $NPSH_a$
➢ Solution: Bring level back in vessel and install safety measures (level sensors) to prevent liquid level dropping below a certain level.

❖ **Air leaks upstream the pump**

➢ Example 1: A normally pressurized suction-side vessel has air leaks moving the absolute pressure on the liquid (H_a) closer towards ambient pressure reducing the $NPSH_a$.

➢ Solution: Leak-testing and fixing.
➢ Example 2: Leaks in piping bends or fittings will result in a lower $NPSH_a$.
➢ Solution: Leak-testing the entire upstream pipe and fixing any leaks.

❖ **Increase in friction upstream the pump**

➢ Example 1: Pipes or fittings on the suction-side of the pumps can be fouled/damaged in such a way that they produce a bigger pressure loss.
➢ Cleaning/repairing/replacing the fouled or damaged piping/fittings.

❖ **Changes in fluid properties**

➢ Example 1: A significant temperature increase of certain fluids can increase the vapor pressure reducing $NPSH_a$.
➢ Solution: Develop an acceptable range of temperature of operation for the fluid, and take steps that the fluids stay within range even during changes in ambient temperature conditions.

❖ **Changes in System Design:**

➢ Example 1: Extra fittings were installed on the suction side of the pump, increasing the pressure loss due to friction, decreasing $NPSH_a$.
➢ Solution: Determine what can be done to increase $NPSH_a$ or reduce $NPSH_r$ without taking out the new fittings (if they are necessary). Could be decreasing the flow rate by increasing friction on the discharge side, reducing the frequency with a VFD to decrease $NPSH_r$. Could also be increasing the liquid surface height of the suction side tank to increase $NPSH_a$. There are several solutions. Depending on the system design change, consider if a new pump is the least expensive work-around long term.

Step 3: Monitor the system performance to check if the corrective actions have been effective.

4.3 Step-by-Step - Decrease Pump Capacity in the Operational Phase

By Operationional Phase it is meant that the pump system is already built and is in operation. Sometimes changes in further downstream processes require a decrease in flow rate by a particular pump. These are steps to take to ensure a safe decrease in pumping capacity.

Step 1: Determine the required decrease in flow rate according to the new operational requirements.

Step 2: Identify the best option in terms of time and cost:

❖ Adjusting the speed (RPM) of the pump with a VFD as discussed in subchapter 3.6, while ensuring that the pump is still operating at a decent efficiency.

❖ Install a smaller impeller on the pump, making sure that pump efficiency is still good and that the material of the new impeller is compatible with the fluid.

❖ Install a throttle valve downstream the pump or partly close an existing valve downstream the pump to increase the head needed to be produced by the pump thereby reducing the flow rate. As seen in the example in subchapter 3.5, this can sometimes be less energy efficient than acquiring a VFD.

❖ Using a bypass line to recirculate some of the discharge flow back to the suction side, thereby limiting the net flow rate. This method can be very costly if no bypass line is currently in place.

❖ Purchasing and implementing a smaller capacity pump with a pump curve better suited for the desired operational requirements. This will involve a significant investment, but could prove to be the better financial decision due to improved efficiency, performance and maintenance costs long term.

Step 3: If one of the four first options are chosen (not acquiring a new pump) make sure to review the OEM's specifications and design parameters on the pump and determine the effect that the proposed change will have on the resulting change to the pump curve or system curve.

Step 4: Leveraging simulation software (like AFT Fathom or Aspen Plus) or manual calculations ensure that the option chosen will lead to operation that matches the desired operational requirement whilst maintaining a healthy relationship between $NPSH_a$ and $NPSH_r$.

Step 5: Implement the change to the system or pump.

Step 6: Do an effectiveness check, carefully monitoring key parameters such as:

- ❖ Flow rate
- ❖ Pressure
- ❖ Power consumption
- ❖ Irregular vibration and noise
- ❖ Wear and tear (visible leakage for instance)

4.4 Step-by-Step - Increase Pump Capacity in the Operational Phase

By operation phase it is meant that the pump system is already built and is in operation. Sometimes changes in further downstream processes require a decrease in flow rate by a particular pump. These are steps to take to ensure a safe decrease in pumping capacity.

Step 1: Determine the required decrease in flow rate according to the new operational requirements. Make sure to check if the current pump installed has the desired increased capacity available according to the pump curve. If not, skip to the last option in Step 2 as a new pump will be needed.

Step 2: Identify the best option in terms of time and cost:

❖ If the pump is already operating at a lower frequency using a VFD, simply increasing the frequency will increase the flow rate.

❖ Removing blockages and unnecessary restrictions in the system. Could for instance be an out-of-service flow meter or valve, or simply accumulated solids decreasing the flow rate.

❖ Installing a bigger impeller on the pump, making sure that pump efficiency is still good and that the material of the new impeller is compatible with the fluid.

❖ Increase the opening% of a valve downstream the pump

❖ Purchasing and implementing a larger capacity pump. While this option might require the highest upfront investment, it might be the best long term option if performance and efficiency at the desired flow rate can be achieved due to power consumption and maintenance costs.

Step 3: If one of the four first options are chosen (not acquiring a new pump) make sure to review the OEM's specifications and design parameters on the pump and determine the effect that the proposed change will have on the resulting change to the pump curve or system curve.

Step 4: Leveraging simulation software (like AFT Fathom or Aspen Plus) or manual calculations ensure that the option chosen will lead to operation that matches the desired operational requirement whilst maintaining a healthy relationship between $NPSH_a$ and $NPSH_r$. This is especially important when looking to increase the flow rate, since this will lead to a higher $NPSH_r$ most of the time, thereby increasing the risk of cavitation.

Step 5: Implement the change to the system or pump.

Step 6: Do an effectiveness check, carefully monitoring key parameters such as:

- ❖ Flow rate
- ❖ Pressure
- ❖ Power consumption
- ❖ Irregular vibration and noise
- ❖ Wear and tear (visible leakage for instance)

5. How to Create Impactful Engineering Project Proposals: A Guide to Effective Value Propositions

In order to become a trusted advisor in your business environment you should hone your skills in creating a persuasive project proposal that effectively communicates the technical and financial aspects of the proposed project to stakeholders and decision-makers. By the end of this chapter you are, hopefully, able to:

❖ Understand the factors that affect the financial viability of an engineering project proposal
❖ Conduct a cost-benefit analysis to evaluate the potential financial impact of a proposed project
❖ Identify and quantify the potential costs, risks and benefits of a proposed project
❖ Develop a financial model to evaluate the economic viability of a proposed project
❖ Understand the time value of money and its impact on investment decisions
❖ Analyze project cash flows and calculate financial metrics such as net present value (NPV), internal rate of return (IRR), and payback period

5.1 What Does a Project Proposal Look Like in The Eye of a Decision Maker?

See yourself as a servant of very busy people who need your expertise without having to understand all of your engineering knowledge.

Most business decisions are made on the basis of an executive summary which provides a quick-read overview over pains and gains. Even in situations when very large investments in technical improvement projects are recommended, keep in mind that many executives will focus only on the financial impact of your proposal and maybe skim through your appendices with your detailed technical and financial calculations.

A note on accountability:

As a young engineer it can be an intimidating step to incorporate the financial aspects of an engineering project into their list of responsibilities, as taking on the extra financial accountability comes with the downside of being the owner of a potentially bad financial decision. While this can sound frightening, the upside is well worth the risk. The upside of taking on this accountability is becoming a non-commodity engineer. When things go well and the projects are financially successful you will also own a significant part of that success, and this type of success is noticed more by founders and key stakeholders than success in classically technical engineering value additions in a company.

When you start taking on more accountability, more resources will naturally flow your way, as most companies are starving for accountable and financially competent engineers. These resources will show up as an increase in your professional mandate, more promotion opportunities and in your bank account.

The following page will show an example of an executive summary, so it is clear what the end product built with the calculations shown later in this chapter is:

Project proposal: Repair or renew critical reactor vessel at X Corp.			
Problem statement:	Observations have shown a downward trend in the performance of reactor vessel [XYZ] since [month/year]. The business is impacted by a yearly loss of yield and increasing risk of production down-time.		
	Scenario 1	**Scenario 2**	**Scenario 3**
Solution approach	Invest in a new reactor vessel	Repair the damaged reactor vessel	Do Nothing
Investment (procurement & implementation)	$1,000,000	$200,000	$0
NPV estimate	$434,706	$355,612	-$84,983
IRR estimate	21%	47%	n.a.
Payback Period	3.3 years	2 years	n.a.
Uncertainty of calculations	Low	Medium	Low
Project risk	Low	Medium	n.a.
Technical recommendation	From an engineering standpoint Scenario 1 is the preferred long-term solution which will reduce technical uncertainties the most. A short-term investment perspective, if necessary, will favor Scenario 2.		

Nevertheless, you should always share your underlying assumptions and calculations. Technical executives might take a deep dive or some business managers might order a second technical opinion. And if your details provided as appendices with the executive summary are shown to hold up to scrutiny, you will earn trust as an advisor.

Trust earned this way compounds heavily over time and the more solid value propositions you execute, the less scrutiny you will experience over time.

5.2 The Fundamentals of Financial Analysis

Process engineers are generally technically gifted and can create valuable engineering analysis and optimize systems to be as efficient and durable as possible.

Some process engineers stand-out among their peers to the whole organization and its managers by having one extra key skill. That is, being able to turn process engineering calculations into a well defined financial analysis of the alternatives moving forward with the technical calculations as a basis.

In order to build sound value proposals for different investments and interventions, some financial definitions that are especially frequent in engineering have to be in the engineer's knowledge bank.

Liquidity is the amount of cash that a company has on hand available to spend on investments.

CAPEX (Capital Expenditure) is the money spent on upgrading, repairing or acquiring physical assets such as VFDs, pumps, valves, vessels, piping etc.

OPEX (Operating Expenses) is the money spent on running the facility on a day-to-day basis, where costs fall in groups like: labor costs, energy costs, raw material and consumables, maintenance and repair, insurance etc.

Discount Rate is a form of interest rate subjective to the given field and market that the company operates in. It is used to calculate the net present value of future changes in cash flow. Money now is worth more than the same amount in a year due to opportunity costs, inflation etc., and this is why it is useful to calculate future costs/revenue to net present value. Most companies use discount rates from 6-12 % to calculate the net present value of their future cash flows. The reason for where the company sets their discount rate is based on how much return they expect that they normally can generate by investing an amount of money now. It reflects the project's risk and opportunity costs.

NPV (Net Present Value) as mentioned above is a method of calculating future increased/decreased costs/revenues to a present day value. A s

$$P = F(1 + i)^{-n}$$

Where P is present value, F is a singular future value, i is the given interest rate per n, and n is the number of periods before the future value is obtained/spent. The sum of all present value calculations (done for changes in costs and revenues) is called Net Present Value (NPV).

Let's take a simplified example where repairing a pump now is estimated to yield a decrease in costs for the next maintenance period (in 1 year from now). The company's discount rate is for this example set at 10 %.

$$P = \$2000(1 + 0.1)^{-1} = \$1818$$

This basically means that having to pay $2000 less during the next maintenance period is worth $1818 dollars to the company now.

Usually changes done to a system will have "perpetual annuities", meaning that there will be a continuous increase in revenue or decrease in costs "forever" in the future as a result of an investment now.

$$P = \frac{A}{i}$$

Where A is the perpetual annuities per time period of which the interest rate i accounts. This is actually used in the real world to calculate perpetual cash flow.

To calculate the net present value of perpetual annuities, let's take an example of restructuring a system of piping and pumps that is estimated to increase yield by 2 %, and thereby increasing yearly revenue by $100,000.

$$P = \frac{\$100{,}000}{0.1} = \$1{,}000{,}000$$

This means that this investment is worth $1 million to the company today. If the costs are less than that, the company is usually willing to do it, and if the costs are higher than that the company should probably not do it.

A critical factor to this is ofcourse the i (interest rate) used in the equation, which is the discount rate. The actual discount rate for many companies can fluctuate because of liquidity issues or abundance during times of recession or financially great times.

If a cost, saving or extra revenue are estimated to happen on a recurring basis but not in perpetuity the following equation for ordinary simple annuities can be used:

$$P = A\left[\frac{1-(1+i)^{-n}}{i}\right]$$

This equation could for example be used to calculate the present value of your netflix subscription if you decide to stay subscribed for the next 48 months. You would then use a discount rate, that is equivalent to the rate of return you believe you could get by investing the money in stocks (for instance). This would reveal the real present costs of your netflix account. In this example this is your present opportunity cost for wanting to watch Netflix.

When looking at a cash flow outcome of an investment (all movement of money - all cash in and out of the company directly related to the investment) all flows have to be calculated back to present value using the three equations above.

Take the sum of all present values (P) of all futures revenues, savings and costs together with all actual present costs for the investment. If the total sum is larger than 0, then the investment is profitable according to the company's discount rate. If the sum is smaller than 0, it is not profitable according to the company's discount rate.

Payback Period is the amount of time it takes for a CAPEX investment to pay for itself with future cash flow (in engineering usually measured from an increase in production output or a decrease in OPEX). Payback Period is also a metric used in value proposals that tell a similar story as NPV, but using less abstract thinking than NPV. Let's calculate the payback period for the previous piping and pump system investment example:

Let's say a quote has been made for the investment and the estimated costs come out to $900,000. It was estimated the revenue would increase with $100,000 per year, meaning that after 9 years the investment would be made back.

Companies with healthy liquidity usually run on NPV-basis for their investment decisions. Some companies for which liquidity is an issue or fluctuates a lot, the payback period can be a very important metric for higher management in deciding between alternatives. Depending on the companies' liquidity status, 9 years might be a good payback period and for some companies it might be way too long. It is therefore usually good practice to present higher management with both NPV- and Payback Period calculations.

Internal Rate of Return (IRR) is also a common measurement of the profitability of an investment. The IRR is the discount rate that would make the NPV of an investment's total cash flow equal to zero (the profitability point). The higher the IRR, the more profitable an investment is.

It is found by setting any of the 3 previous present value calculations equal to zero and solving for the discount rate. If the calculated IRR is higher than the normal discount rate for the company it is a profitable investment - basically meaning this investment will outperform the baseline interest that the company gains through the average investment of cash.

A Caveat about the suggested financial definitions:

Make sure your financial definitions are agreed by stakeholders. Unless the definitions for CAPEX, OPEX, Discount Rate and NPV are perfectly ready and consensual in your business environment, you should consult your accounting / controller function the first time you make an important project proposal.

Definitions may vary a lot between companies and industries and even internally between business divisions. Sometimes procurement and implementation of physical equipment is depreciated as CAPEX and sometimes depending on price or tradition such costs are accounted for as OPEX.

If you learn to use the business definitions listed here you will be able to live up to comprehensive business case requirements in most companies. But please note that using all these definitions might be overkill in some instances. Sometimes a much simpler cost-benefit analysis is required, and your business executive or your technical manager will be perfectly happy with an overview of initial cost and future savings all calculated in present value in rough estimates.

Accordingly, it is recommended to set expectations with the users of your project proposal, e.g., your manager or your business client.

The following subchapter will show how to derive the internal calculations needed to craft a project proposal similar to the one presented in subchapter 5.1.

5.3 Developing a Value Proposal

Enough theory, let's go through the mathematical content of the executive summary from subchapter 5.1. Three alternatives for fixing a slightly compromised biogas reactor vessel that is deteriorating with time (the company's internal discount rate assumed at 10 %) has been identified:

Alternative 1: Invest in a new reactor vessel
Investment: $1,000,000 (all investment/implementation costs included)
Expected increased cash flow due to better cold gas efficiency:

- Year 1-5: $300,000/year
- Year 6-10: $100,000/year

1. Calculating **NPV** using the equation for ordinary simple annuities:

$$NPV = \$300,000\left[\frac{1-(1+0.1)^{-5}}{0.1}\right] + \$100,000\left[\frac{1-(1+0.1)^{-10}}{0.1}\right]$$
$$- \$100,000\left[\frac{1-(1+0.1)^{-5}}{0.1}\right] - \$1,000,000$$
$$NPV = \$1,137,236 + \$614,456 - \$316,986 - \$1,000,000$$
$$NPV = \$434,706$$

2. Finding the **IRR** by solving for the discount rate where NPV = 0:

$$0 = \$300,000\left[\frac{1-(1+i)^{-5}}{i}\right] + \$100,000\left[\frac{1-(1+0.1)^{-10}}{i}\right]$$
$$- \$100,000\left[\frac{1-(1+0.1)^{-5}}{i}\right] - \$1,000,000$$

$$i = IRR = 0.21 = 21\%$$

3. Calculating the **Payback Period**:

$$0 = \$300,000/year \times years - \$1,000,000$$
$$years = \frac{\$1,000,000}{\$300,000/year} = 3.3\ years$$

Alternative 2: Repair the damage reactor vessel

Investment: $200,000 (all investment/implementation costs included)
Expected increased cash flow due to better cold gas efficiency:

- Year 1-5: $100,000/year
- Year 6-10: $75,000/year

1. Calculating **NPV** using the equation for ordinary simple annuities:

$$NPV = \$100,000\left[\frac{1-(1+0.1)^{-5}}{0.1}\right] + \$75,000\left[\frac{1-(1+0.1)^{-10}}{0.1}\right]$$
$$- \$75,000\left[\frac{1-(1+0.1)^{-5}}{0.1}\right] - \$200,000$$

$$NPV = \$355,612$$

2. Finding the **IRR** by solving for the discount rate where NPV = 0:

$$0 = \$100,000\left[\frac{1-(1+i)^{-5}}{i}\right] + \$75,000\left[\frac{1-(1+0.1)^{-10}}{i}\right]$$
$$- \$75,000\left[\frac{1-(1+0.1)^{-5}}{i}\right] - \$200,000$$

$$i = IRR = 0.47 = 47\%$$

3. Calculating the **Payback Period**:

$$0 = \$100,000/year \times years - \$200,000$$

$$years = \frac{\$200,000}{\$100,000/year} = 2\ years$$

Alternative 3: Do nothing.
Investment: $0

Expected **decreased** cash flow due to poorer cold gas efficiency:

- Year 1-5: -$10,000/year
- Year 6-10: -$20,000/year

$$NPV = - \$10,000 \left[\frac{1-(1+0.1)^{-5}}{0.1} \right] - \$20,000 \left[\frac{1-(1+0.1)^{-10}}{0.1} \right]$$
$$+ \$20,000 \left[\frac{1-(1+0.1)^{-5}}{0.1} \right]$$

$$NPV = -\$84,983$$

A negative NPV, can then be used as a cost term, when calculating a different alternative of using the money - for instance using the liquidity of the company on another investment for a new gas engine or boiler. The more alternatives the process engineer can show the better the limited liquidity of the company can be allocated to maximize their IRR.

This negative NPV is the present cost of not doing anything about the deteriorating reactor vessel and should be included in the calculation when allocating liquid capital towards other investment opportunities.

It is important to notice that this is a simplified example. A real value proposition calculation like this one should include all potential changes in cash flow (revenue, costs). In this example, the changes in scheduled recurring or unscheduled maintenance costs are for instance omitted from the calculations.

The calculations for NPV, IRR and payback period are not necessary to show if presenting the alternatives to executive management - have them handy in an appendix instead. However, for presenting your results to the maintenance manager/facility manager or similar level supervisor it might be a good idea to present the calculations. It depends on the situation, their awareness of the issue/data and many other factors whether or not you want the calculations in the proposal or not, but always have them ready as an appendix.

The financially competent engineer's job in a value proposal situation is to present key figures concisely and differently depending on the receiver. The value proposal should look a certain way if presented to your immediate boss and a different way if presented to executive managers, board members and stakeholders.

An important note of human psychology:

The process engineer should present all the alternatives discovered but not conclude too much on the findings. Executive managers will usually prefer to be (or rather look like) the smartest person in the room by choosing the best options from the presented alternatives. Be careful of outshining the master by jumping to your own conclusions.

6. Conclusion

Thanks for reading!

Whether you read through the entire book or are using it as a reference tool in your work at university or in your professional engineering endeavors, you have explored some of the most prominent and value-adding process engineering concepts. There will always be a need for good engineers to design pumps and hydraulic systems, and the engineers that can turn that knowledge into financial advice of where a company should allocate their money will forever be in exceptionally high demand.

By following the practice examples and step-by-step guides, you have hopefully broadened your mathematical understanding of pump calculations and financial analysis and have learned how to apply them to solve real-world engineering problems in a professional and persuasive manner.

My hope is that the knowledge obtained from this book will empower you to approach complex engineering problems with confidence and provide you with a solid foundation for further study and work in the field of process engineering.

May you find purpose, fulfillment, and joy in everything that you create.
Michael Kay Hoffmann

7. Appendices

Appendix 7.1: Roughness Coefficients of Common Piping Materials

Surface	Absolute Roughness Coefficient ε	
	$(10\text{-}3\text{ m}) = \text{mm}$	(feet)
Drawn Copper, Lead, Brass, Aluminum (new)	0.001 - 0.002	(3.28 - 6.56) 10-6
PVC, PE and other smooth Plastic Pipes	0.0015 - 0.007	(0.49 - 2.30) 10-5
Stainless steel, bead blasted	0.001 - 0.006	(0.00328 - 0.0197) 10-3
Stainless steel, turned	0.0004 - 0.006	(0.00131 - 0.0197) 10-3
Stainless steel, electron-polished	0.0001 - 0.0008	(0.000328 - 0.00262) 10-3
Commercial steel or wrought iron	0.045 - 0.09	(1.48 - 2.95) 10-4
Stretched steel	0.015	4.95 10-5
Weld steel	0.045	1.48 10-4
Galvanized steel	0.15	4.92 10-4
Rusted steel (corrosion)	0.15 - 4	(4.92 - 131) 10-4
New cast iron	0.25 - 0.8	(8.2 - 26.2) 10-4
Worn cast iron	0.8 - 1.5	(2.62 - 4.92) 10-3
Rusty cast iron	1.5 - 2.5	(4.92 - 8.2) 10-3
Sheet or asphalted cast iron	0.01 - 0.015	(3.28 - 4.92) 10-5
Smoothed cement	0.3	0.98 10-3

Values from: https://www.engineeringtoolbox.com/ (14APR2023)

Appendix 7.2: Common Fittings and Their Equivalent Pipe Lengths

Fittings	Rigid PVC/HDPE e = 0.005 mm	GRP/FRP e = 0.02 mm	Commercial Steel e = 0.05 mm	Spiral Weld Steel e = 0.1 mm
Threaded bends				
90o elbow, r/d=1	37	34	30	26
45o elbow, r/d=1	20	18	16	14
90o elbow, sharp bend	69	63	55	49
90o elbow, r/d=1	23	21	19	16
90o elbow, r/d=1.5	17	15	13	12
90o elbow, r/d=2	14	13	11	10
45o elbow, sharp bend	22	20	18	16
45o elbow, r/d=1	17	16	14	12
45o elbow, r/d=1.5	12	11	9.4	8.3
Threaded tees				
Tee, straight through	25	23	20	18
Tee, through branch	75	68	60	53
Welded tees				
Tee, square, straight through	0	0	0	0
Tee, square, through branch	87	79	70	61
Tee, radiused, straight through	13	12	10	9
Tee, radiused, through branch	72	65	57	50
Valves / Strainers				

Fittings	Rigid PVC/HDPE e = 0.005 mm	GRP/FRP e = 0.02 mm	Commercial Steel e = 0.05 mm	Spiral Weld Steel e = 0.1 mm
Globe valve, full open	400	370	320	280
Gate valve, full open	9	8.5	7.5	6.6
Ball valve, full bore	3.3	3.0	2.6	2.3
Ball valve, reduced bore	31	28	25	22
Plug valve, 2-way	21	19	17	15
Plug valve, 3-way, straight through	36	32	29	25
Plug valve, 3-way, through branch	100	95	84	74
Diaphragm valve, weir type	200	190	160	140
Butterfly valve	46	42	37	32
Lift check valve	700	640	560	490
Swing check valve	120	110	95	85
Wafer disk check valve	530	480	420	370
Y-strainer, clean	300	280	250	220

https://www.katmarsoftware.com/

Appendix 7.3: Suggested Reading

As this book functions and reads like a handbook, further reading on the subjects of this book is recommended for a deeper understanding of the concepts presented. The books below are helpful for this purpose.

"Fluid Mechanics" by Frank White - This book provides a comprehensive introduction to fluid mechanics and is an excellent resource for understanding the principles behind hydraulic systems.

"Centrifugal Pumps: Design and Application" by John Tuzson - For readers interested in a more detailed understanding of centrifugal pumps, this book provides a comprehensive overview of design principles and applications.

"Engineering Economic Analysis" by Don Newnan, Ted Eschenbach, Jerome Lavelle, Neal Lewis- For readers looking to deepen their understanding of financial analysis, this book provides a thorough understanding of engineering economics and financial decision making.

"Project Management for Engineering, Business and Technology" John M. Nicholas, Herman Steyn - This book provides a comprehensive overview of project management principles and best practices specifically tailored for engineering projects.

"Influence: The Psychology of Persuasion" by Robert Cialdini - For readers interested in developing their persuasion skills, this book provides a comprehensive overview of the psychology of influence and practical techniques for achieving desired outcomes.

www.ingramcontent.com/pod-product-compliance
Lightning Source LLC
Chambersburg PA
CBHW072032230526
45466CB00020B/1786